Springer Undergraduate Mathematics Series

Springer
London
Berlin
Heidelberg
New York
Barcelona
Hong Kong
Milan
Paris
Santa Clara
Singapore
Tokyo

Other books in this series

Analytic Methods for Partial Differential Equations *G. Evans, J. Blackledge, P. Yardley*
Basic Linear Algebra *T.S. Blyth and E.F. Robertson*
Basic Stochastic Processes *Z. Brzeźniak and T. Zastawniak*
Elements of Logic via Numbers and Sets *D.L. Johnson*
Elementary Number Theory *G.A. Jones and J.M. Jones*
Groups, Rings and Fields *D.A.R. Wallace*
Hyperbolic Geometry *J.W. Anderson*
Introduction to Laplace Transforms and Fourier Series *P.P.G. Dyke*
Introduction to Ring Theory *P.M. Cohn*
Introductory Mathematics: Algebra and Analysis *G. Smith*
Introductory Mathematics: Applications and Methods *G.S. Marshall*
Measure, Integral and Probability *M. Capińksi and E. Kopp*
Multivariate Calculus and Geometry *S. Dineen*
Numerical Methods for Partial Differential Equations *G. Evans, J. Blackledge, P. Yardley*
Topologies and Uniformities *I.M. James*
Vector Calculus *P.C. Matthews*

James W. Anderson

Hyperbolic Geometry

With 20 Figures

Springer

James W. Anderson, PhD
Faculty of Mathematical Studies, University of Southampton, Highfield, Southampton SO17 1BJ, UK

Cover illustration elements reproduced by kind permission of:

Aptech Systems, Inc., Publishers of the GAUSS Mathematical and Statistical System, 23804 S.E. Kent-Kangley Road, Maple Valley, WA 98038, USA. Tel: (206) 432 - 7855 Fax (206) 432 - 7832 email: info@aptech.com URL: www.aptech.com

American Statistical Association: Chance Vol 8 No 1, 1995 article by KS and KW Heiner 'Tree Rings of the Northern Shawangunks' page 32 fig 2

Springer-Verlag: Mathematica in Education and Research Vol 4 Issue 3 1995 article by Roman E Maeder, Beatrice Amrhein and Oliver Gloor 'Illustrated Mathematics: Visualization of Mathematical Objects' page 9 fig 11, originally published as a CD ROM 'Illustrated Mathematics' by TELOS: ISBN 0-387-14222-3, German edition by Birkhauser: ISBN 3-7643-5100-4.

Mathematica in Education and Research Vol 4 Issue 3 1995 article by Richard J Gaylord and Kazume Nishidate 'Traffic Engineering with Cellular Automata' page 35 fig 2. Mathematica in Education and Research Vol 5 Issue 2 1996 article by Michael Trott 'The Implicitization of a Trefoil Knot' page 14.

Mathematica in Education and Research Vol 5 Issue 2 1996 article by Lee de Cola 'Coins, Trees, Bars and Bells: Simulation of the Binomial Process page 19 fig 3. Mathematica in Education and Research Vol 5 Issue 2 1996 article by Richard Gaylord and Kazume Nishidate 'Contagious Spreading' page 33 fig 1. Mathematica in Education and Research Vol 5 Issue 2 1996 article by Joe Buhler and Stan Wagon 'Secrets of the Madelung Constant' page 50 fig 1.

ISBN 1-85233-156-9 Springer-Verlag London Berlin Heidelberg

British Library Cataloguing in Publication Data
Anderson, James W.
Hyperbolic geometry. - (Springer undergraduate mathematics series)
1. Geometry, Hyperbolic
I. Title
516.9
ISBN 1852331569

Library of Congress Cataloging-in-Publication Data
Anderson, James W., 1964-
Hyperbolic geometry / James W. Anderson.
p. cm. -- (Springer undergraduate mathematics series)
Includes bibliographical references and index.
ISBN 1-85233-156-9 (alk. paper)
1. Geometry, Hyperbolic. I. Title. II. Series.
QA685.A54 1999 99-37719
516.9—dc21 CIP

Typesetting: Camera ready by author
Printed and bound at the Athenæum Press Ltd., Gateshead, Tyne & Wear
12/3830-543210 Printed on acid-free paper SPIN 10682618

Contents

Preamble

What you have in your hands is an introduction to the basics of planar hyperbolic geometry. Writing this book was difficult, not because I was at any point at a loss for topics to include, but rather because I continued to come across topics that I felt should be included in an introductory text. I believe that what has emerged from the process of writing gives a good feel for the geometry of the hyperbolic plane.

This book is written to be used either as a classroom text or as more of a self-study book, perhaps as part of a directed reading course. For that reason, I have included solutions to all the exercises. I have tried to choose the exercises to give reasonable coverage to the sorts of calculations and proofs that inhabit the subject. The reader should feel free to make up their own exercises, both proofs and calculations, and to make use of other sources.

I have also tried to keep the exposition as self-contained as possible, and to make as little use of mathematical machinery as possible. The book is written for a third or fourth year student who has encountered some Calculus, particularly the definition of arc-length, integration over regions in Euclidean space, and the change of variables theorem; some Analysis, particularly continuity, open and closed sets in the plane, and infimum and supremum; has a familiarity with Complex Numbers, as most of the book takes place in the complex plane $\mathbb{C}$, but need not have taken a class in Complex Analysis; and some Abstract Algebra, as we make use of some of the very basics from the theory of groups.

Non-Euclidean geometry in general, and hyperbolic geometry in particular, is an area of mathematics which has an interesting history and which is still being actively studied by researchers around the world. One reason for the continuing interest in hyperbolic geometry is that it touches on a number of different fields, including but not limited to Complex Analysis, Abstract Algebra and Group Theory, Number Theory, Differential Geometry, and Low-dimensional Topology.

This book is not written as an encyclopedic introduction to hyperbolic geometry but instead offers a single perspective. Specifically, I wanted to write a hyperbolic geometry book in which very little was assumed, and as much as possible was derived from following Klein's view that geometry, in this case hyperbolic geometry, consists of the study of those quantities invariant under a group. Consequently, I did not want to write down, without what I felt to be reasonable justification, the hyperbolic element of arc-length, or the group of hyperbolic isometries, but instead wanted them to arise as naturally as possible. And I think I have done that in this book.

There is a large number of topics I have chosen not to include, such as the hyperboloid and Klein models of the hyperbolic plane. Also, I have included nothing of the history of hyperbolic geometry and I have not taken the axiomatic approach to define the hyperbolic plane. One reason for these omissions is that there are already a number of excellent books on both the history of hyperbolic geometry and on the axiomatic approach, and I felt that I would not be able to add anything of note to what has already been done. There is an extensive literature on hyperbolic geometry. The interested reader is directed to the list of sources for Further Reading at the end of the book.

And now, a brief outline of the approach taken in this book. We first develop a model of the hyperbolic plane, namely the upper half-plane model $\mathbb{H}$, and define what we mean by a hyperbolic line in $\mathbb{H}$. We then try to determine a reasonable group of transformations of $\mathbb{H}$ that takes hyperbolic lines to hyperbolic lines, which leads us to spend some time studying the group Möb^+ of Möbius transformations and the general Möbius group Möb.

After determining the subgroup $\text{Möb}(\mathbb{H})$ of Möb preserving $\mathbb{H}$, we derive an invariant element of arc-length on $\mathbb{H}$. That is, we derive a means of calculating the hyperbolic length of a path $f : [a, b] \to \mathbb{H}$ in such a way that the hyperbolic length of a path is invariant under the action of $\text{Möb}(\mathbb{H})$, which is to say that the hyperbolic length of a path $f : [a, b] \to \mathbb{H}$ is equal to the hyperbolic length of its translate $\gamma \circ f : [a, b] \to \mathbb{H}$ for any element γ of $\text{Möb}(\mathbb{H})$. We are then able to define a natural metric on $\mathbb{H}$ in terms of the shortest hyperbolic length of a path joining a pair of points.

After exploring calculations of hyperbolic length, we move onto a discussion of convexity and of hyperbolic polygons, and then to the trigonometry of polygons in the hyperbolic plane and the three basic laws of trigonometry in the hyperbolic plane. We also determine how to calculate hyperbolic area, and state and prove the Gauss-Bonnet formula for hyperbolic polygons, which gives the hyperbolic area of a hyperbolic polygon in terms of its interior angles. In the course of this analysis, we introduce other models of the hyperbolic plane, particularly the Poincaré disc model $\mathbb{D}$. We close by describing and exploring very briefly what it means for a subgroup of Möb($\mathbb{H}$) to be well-behaved.

I would like to close this introduction with some acknowledgements. I would like to start by thanking Susan Hezlet for suggesting that I write this book, and David Ireland, who watched over its completion. Part of the writing of this book was done while I was visiting the Mathematics Department of Rice University during the 1998-1999 academic year, and I offer my thanks to the department there, particularly Frank Jones, who was chairman at the time and helped arrange my visit. This book is based on lectures from a class on hyperbolic geometry at the University of Southampton in the Fall terms of the 1996-97 and 1997-98 academic years, and I would like to thank the students in those classes, as well as the students at Rice whose sharp eyes helped in the final clean up of the text. The errors that remain are mine.

I would also like to thank all my mathematics teachers from over the years, particularly Ted Shifrin and Bernie Maskit; my parents, Wyatt and Margaret, and my sisters, Elizabeth and Karen, for all their love and support over the years; and to Barbara, who put up with me through the final stages of the writing.

1
The Basic Spaces

In this first chapter, we set the stage for what is to come. Namely, we define the *upper half-plane model* $\mathbb{H}$ of the hyperbolic plane, which is where most of the action of this book takes place. We define *hyperbolic lines* and talk a bit about *parallelism*. In order to aid our construction of a reasonable group of transformations of $\mathbb{H}$, we expand our horizons to consider the *Riemann sphere* $\overline{\mathbb{C}}$, and close the chapter by considering how $\mathbb{H}$ sits as a subset of $\overline{\mathbb{C}}$.

1.1 A Model for the Hyperbolic Plane

We begin our investigation by describing a model of the hyperbolic plane. By a *model*, we mean a choice of an underlying space and a choice of how to represent basic geometric objects, such as points and lines, in this underlying space. As we shall see over the course of the book, there are a large number of possible models for the hyperbolic plane. In order to give as concrete a description of its geometry as possible, we begin by working in a single specific model.

The model of the hyperbolic plane we work in is the *upper half-plane* model. The underlying space of this model is the upper half-plane $\mathbb{H}$ in the complex plane $\mathbb{C}$, defined to be

$$\mathbb{H} = \{z \in \mathbb{C} : \text{Im}(z) > 0\}.$$

We use the usual notion of point that $\mathbb{H}$ inherits from $\mathbb{C}$. We also use the usual notion of angle that $\mathbb{H}$ inherits from $\mathbb{C}$; that is, the angle between two curves in $\mathbb{H}$ is defined to be the angle between the curves when they are considered to be curves in $\mathbb{C}$, which in turn is defined to be the angle between their tangent lines.

As we will define hyperbolic lines in $\mathbb{H}$ in terms of Euclidean lines and Euclidean circles in $\mathbb{C}$, we begin with a couple of calculations in $\mathbb{C}$.

Exercise 1.1

Express the equations of the Euclidean line $ax + by + c = 0$ and the Euclidean circle $(x-h)^2+(y-k)^2 = r^2$ in terms of the complex coordinate $z = x + yi$ in $\mathbb{C}$.

Exercise 1.2

Let $\mathbb{S}^1 = \{z \in \mathbb{C} \mid |z| = 1\}$ be the unit circle in $\mathbb{C}$. Let A be a Euclidean circle in $\mathbb{C}$ with Euclidean centre $re^{i\theta}$, $r > 1$, and Euclidean radius $s > 0$. Show that A is perpendicular to $\mathbb{S}^1$ if and only if $s = \sqrt{r^2 - 1}$.

We are now ready to define a *hyperbolic line* in $\mathbb{H}$.

Definition 1.1

There are two seemingly different types of *hyperbolic line*, both defined in terms of Euclidean objects in $\mathbb{C}$. One is the intersection of $\mathbb{H}$ with a Euclidean line in $\mathbb{C}$ perpendicular to the real axis $\mathbb{R}$ in $\mathbb{C}$. The other is the intersection of $\mathbb{H}$ with a Euclidean circle centred on the real axis $\mathbb{R}$.

Some examples of hyperbolic lines in $\mathbb{H}$ are shown in Fig. 1.1.

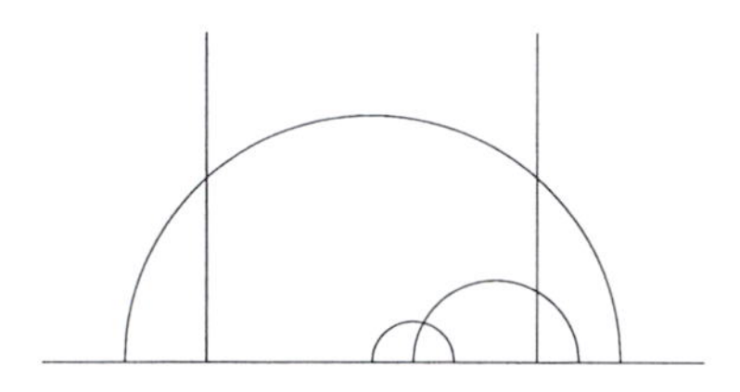

Figure 1.1: Hyperbolic lines in $\mathbb{H}$

We will see in Section 1.2 a way of unifying these two different types of hyperbolic line. For the moment, though, we content outselves with an exploration of some of the basic properties of hyperbolic geometry with this definition of hyperbolic line.

Working in analogy with what we know from Euclidean geometry, there is one property that hyperbolic lines in $\mathbb{H}$ should have, namely that there should always exist one and only one hyperbolic line between any pair of distinct points of $\mathbb{H}$. That this property holds in $\mathbb{H}$ with hyperbolic lines as defined above is a fairly straightforward calculation.

Proposition 1.2

For each pair p and q of distinct points in $\mathbb{H}$, there exists a unique hyperbolic line ℓ in $\mathbb{H}$ passing through p and q.

There are two cases to consider. Suppose first that $\mathrm{Re}(p) = \mathrm{Re}(q)$. Then, the Euclidean line L given by the equation $L = \{z \in \mathbb{C} \mid \mathrm{Re}(z) = \mathrm{Re}(p)\}$ is perpendicular to the real axis and passes through both p and q. So, the hyperbolic line $\ell = \mathbb{H} \cap L$ is the desired hyperbolic line through p and q.

Suppose now that $\mathrm{Re}(p) \neq \mathrm{Re}(q)$. Since the Euclidean line through p and q is no longer perpendicular to $\mathbb{R}$, we need to construct a Euclidean circle centred on the real axis $\mathbb{R}$ that passes though p and q.

Let L_{pq} be the Euclidean line segment joining p and q and let K be the perpendicular bisector of L_{pq}. Then, every Euclidean circle that passes through p and q has its centre on K. Since p and q have non-equal real parts, the Euclidean line K is not parallel to $\mathbb{R}$, and so K and $\mathbb{R}$ intersect at a unique point c.

Let A be the Euclidean circle centred at this point of intersection c with radius $|c-p|$, so that A passes through p. Since c lies on K, we have that $|c-p| = |c-q|$, and so A passes through q as well. The intersection $\ell = \mathbb{H} \cap A$ is then the desired hyperbolic line passing through p and q.

The uniqueness of the hyperbolic line passing through p and q comes from the uniqueness of the Euclidean lines and Euclidean circles used in its construction. This completes the proof of Proposition 1.2.

We note here that the argument used to prove Proposition 1.2 actually contains more information. For any pair of distinct points p and q in $\mathbb{C}$ with non-equal real parts, there exists a unique Euclidean circle centred on $\mathbb{R}$ passing through p and q. The crucial point is that the centre of any Euclidean circle passing through p and q lies on the perpendicular bisector K of the Euclidean line segment L_{pq} joining p and q, and K is not parallel to $\mathbb{R}$.

Since we have chosen the underlying space $\mathbb{H}$ for this model of the hyperbolic plane to be contained in $\mathbb{C}$, and since we have chosen to define hyperbolic lines in $\mathbb{H}$ in terms of Euclidean lines and Euclidean circles in $\mathbb{C}$, we are able to use whatever facts about Euclidean lines and Euclidean circles we know to analyze the behaviour of hyperbolic lines. We have in effect given ourselves familiar coordinates on $\mathbb{H}$ to work with.

For instance, if ℓ is the hyperbolic line in $\mathbb{H}$ passing through p and q, we are able to express ℓ explicitly in terms of p and q. When p and q have equal real parts, we have already seen that $\ell = \mathbb{H} \cap L$, where L is the Euclidean line $L = \{z \in \mathbb{C} \mid \text{Re}(z) = \text{Re}(p)\}$. The expression of ℓ in terms of p and q in the case that $\text{Re}(p) \neq \text{Re}(q)$ is left as an exercise.

Exercise 1.3

Let p and q be distinct points in $\mathbb{C}$ with non-equal real parts and let A be the Euclidean circle centred on $\mathbb{R}$ and passing through p and q. Express the Euclidean centre c and the Euclidean radius r of A in terms of $\text{Re}(p)$, $\text{Im}(p)$, $\text{Re}(q)$, and $\text{Im}(q)$.

A legitimate question to raise at this point is whether hyperbolic geometry in $\mathbb{H}$, with this definition of hyperbolic line, is actually different from the usual Euclidean geometry in $\mathbb{C}$ we are accustomed to. The answer to this question is an emphatic Yes, hyperbolic geometry in $\mathbb{H}$ behaves very differently from Euclidean geometry in $\mathbb{C}$.

One way to see this difference is to consider the behaviour of parallel lines. Recall that Euclidean lines in $\mathbb{C}$ are parallel if and only if they are disjoint, and we adopt this definition in the hyperbolic plane as well.

Definition 1.3

Two hyperbolic lines are *parallel* if they are disjoint.

In Euclidean geometry, parallel lines exist, and in fact, if L is a Euclidean line and if a is a point in $\mathbb{C}$ not on L, then there exists one and only one line K through a that is parallel to L.

In fact, in Euclidean geometry parallel lines are also equidistant, that is, if L and K are parallel Euclidean lines and if a and b are points on L, then the Euclidean distance from a to K is equal to the Euclidean distance from b to K.

In hyperbolic geometry, parallelism behaves much differently. Though we do not yet have a means of measuring hyperbolic distance, we can consider parallel hyperbolic lines qualitatively.

Theorem 1.4

Let ℓ be a hyperbolic line in $\mathbb{H}$ and let p be a point in $\mathbb{H}$ not on ℓ. Then, there exist infinitely many different hyperbolic lines through p that are parallel to ℓ.

As in the proof of Proposition 1.2, there are two cases to consider. First, suppose that ℓ is contained in a Euclidean line L. Since p is not on L, there exists a Euclidean line K through p that is parallel to L. Since L is perpendicular to $\mathbb{R}$, we have that K is perpendicular to $\mathbb{R}$ as well. So, one hyperbolic line in $\mathbb{H}$ through p and parallel to ℓ is the intersection $\mathbb{H} \cap K$.

To construct another hyperbolic line through p and parallel to ℓ, take a point x on $\mathbb{R}$ between K and L, and let A be the Euclidean circle centred on $\mathbb{R}$ that passes through x and p. We know that such a Euclidean circle A exists since $\mathrm{Re}(x) \neq \mathrm{Re}(p)$.

By construction, A is disjoint from L, and so the hyperbolic line $\mathbb{H} \cap A$ is disjoint from ℓ. That is, $\mathbb{H} \cap A$ is a second hyperbolic line through p that is parallel to ℓ. Since there are infinitely many points on $\mathbb{R}$ between K and L, this construction gives infinitely many different hyperbolic lines through p and parallel to ℓ. A picture of this phenomenon is given in Fig. 1.2.

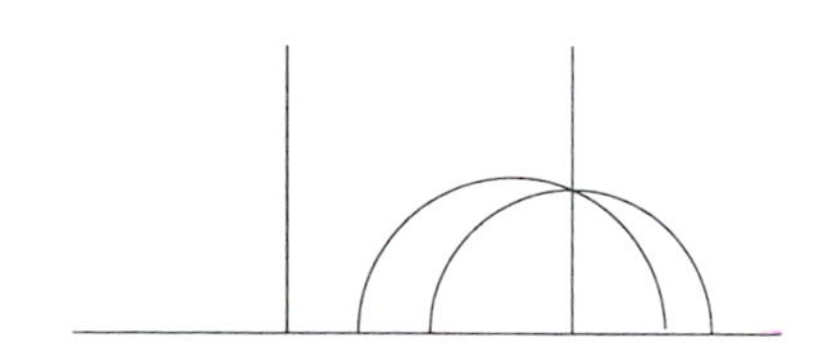

Figure 1.2: Several parallel hyperbolic lines

Exercise 1.4

Give an explicit description of two hyperbolic lines in $\mathbb{H}$ through i and parallel to the hyperbolic line $\ell = \mathbb{H} \cap \{z \in \mathbb{C} \mid \mathrm{Re}(z) = 3\}$.

Now, suppose that ℓ is contained in a Euclidean circle A. Let D be the Euclidean circle that is concentric to A and that passes through p. Since concentric circles are disjoint and have the same centre, one hyperbolic line through p and parallel to ℓ is the intersection $\mathbb{H} \cap D$.

To construct a second hyperbolic line through p and parallel to ℓ, take any point x on $\mathbb{R}$ between A and D. Let E be the Euclidean circle centred on $\mathbb{R}$ that passes through x and p. Again by construction, E and A are disjoint, and so $\mathbb{H} \cap E$ is a hyperbolic line through p parallel to ℓ.

As above, since there are infinitely many points on $\mathbb{R}$ between A and D, there are infinitely many hyperbolic lines through p parallel to ℓ. A picture of this phenomenon is given in Fig. 1.3.

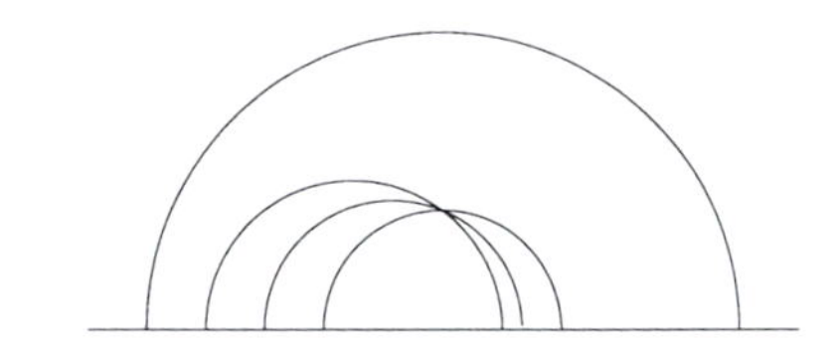

Figure 1.3: Several parallel hyperbolic lines

Exercise 1.5

Give an explicit description of two hyperbolic lines in $\mathbb{H}$ through i and parallel to the hyperbolic line $\ell = \mathbb{H} \cap A$, where A is the Euclidean circle with Euclidean centre -2 and Euclidean radius 1.

We now have a model to play with. The bulk of this book is spent exploring this particular model of the hyperbolic plane, though we do spend some time developing and exploring other models as well.

We close this section with a few words to put what we are doing in a historical context. We are proceeding almost completely backwards in our development of hyperbolic geometry from the historical development of the subject. A much more common approach is to begin with the axiomization of Euclidean geometry. One of the axioms is the statement about parallel lines mentioned above, namely that given a line Euclidean L and a point p not on L, there exists a unique Euclidean line through p and parallel to L. This axiom is often referred to as the Parallel Postulate; the form we give here is credited to Playfair.

Hyperbolic geometry is then defined by using the same set of axioms as Euclidean geometry, with the hyperbolic variant of the Parallel Postulate, namely that given a hyperbolic line ℓ and a point p not on ℓ, there are at least two hyperbolic lines through p and parallel to ℓ.

It is then shown that the upper half-plane model, with hyperbolic lines as we have defined them, is a model of the resulting non-Euclidean geometry. For instance, see the books of Stahl [24] and Greenberg [10], as well as the other sources mentioned in the list of Further Reading.

In this book, we are less concerned with the axiomatic approach to hyperbolic geometry, preferring to make use of the fact that we have reasonable coordinates in the upper half-plane $\mathbb{H}$, which allow us to calculate fairly directly.

Our first major task is to determine whether we have enough information in this definition of hyperbolic geometry to define the notions of hyperbolic length, hyperbolic distance, and hyperbolic area in $\mathbb{H}$. We do this using the group of transformations of $\mathbb{H}$ taking hyperbolic lines to hyperbolic lines.

1.2 The Riemann Sphere $\overline{\mathbb{C}}$

In order to determine the transformations of $\mathbb{H}$ that take hyperbolic lines to hyperbolic lines, we first fulfil our earlier promise of unifying the two seemingly different types of hyperbolic line, namely those contained in a Euclidean line and those contained in a Euclidean circle. We take as our stepping off point the observation that a Euclidean circle can be obtained from a Euclidean line by adding a single point.

To be explicit, let $\mathbb{S}^1$ be the unit circle in $\mathbb{C}$, and consider the function

$$\xi : \mathbb{S}^1 - \{i\} \to \mathbb{R}$$

defined as follows: given a point z in $\mathbb{S}^1 - \{i\}$, let K_z be the Euclidean line passing through i and z, and set $\xi(z) = \mathbb{R} \cap K_z$. This function is well-defined, since K_z and $\mathbb{R}$ intersect in a unique point as long as $\mathrm{Im}(z) \neq 1$. See Fig. 1.4.

This operation is referred to as *stereographic projection*. In terms of the usual cartesian coordinates on the plane, the real axis $\mathbb{R}$ in $\mathbb{C}$ corresponds to the x-axis, and so $\xi(z)$ is the x-intercept of K_z. Calculating, we see that K_z has slope

$$m = \frac{\mathrm{Im}(z) - 1}{\mathrm{Re}(z)}$$

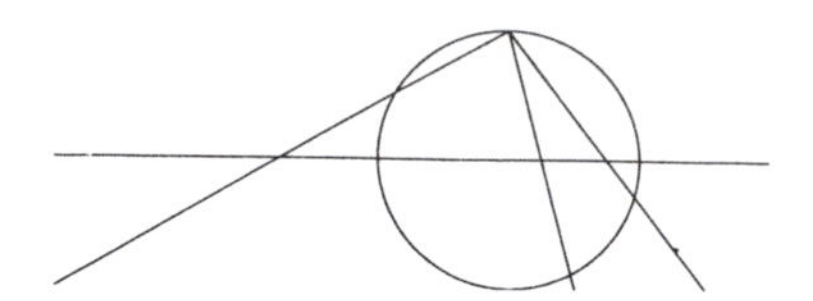

Figure 1.4: Stereographic projection

and y-intercept 1. Hence, the equation for K_z is

$$y - 1 = \frac{\mathrm{Im}(z) - 1}{\mathrm{Re}(z)} x.$$

In particular, the x-intercept of K_z is

$$\xi(z) = \frac{\mathrm{Re}(z)}{1 - \mathrm{Im}(z)}.$$

Exercise 1.6

Give an explicit formula for $\xi^{-1} : \mathbb{R} \to \mathbb{S}^1 - \{i\}$.

Exercise 1.7

Consider the three points $z_k = \exp\left(\frac{2\pi k}{3} i\right)$, $0 \leq k \leq 2$, of $\mathbb{S}^1$ that form the vertices of an equilateral triangle in $\mathbb{C}$. Calculate their images under ξ.

In fact, ξ is a bijection between $\mathbb{S}^1 - \{i\}$ and $\mathbb{R}$. Geometrically, this follows from the fact that a pair of distinct points in $\mathbb{C}$ determines a unique Euclidean line. If z and w are points of $\mathbb{S}^1 - \{i\}$ for which $\xi(z) = \xi(w)$, then K_z and K_w both pass through the same point of $\mathbb{R}$, namely $\xi(z) = \xi(w)$. However, since both K_z and K_w pass through i as well, this forces the two lines K_z and K_w to be equal, and so $z = w$.

Since we obtain $\mathbb{R}$ from $\mathbb{S}^1$ by removing a single point of $\mathbb{S}^1$, namely i, we can think of constructing the Euclidean circle $\mathbb{S}^1$ by starting with the Euclidean line $\mathbb{R}$ and adding a single point.

Motivated by this, one possibility for a space that contains $\mathbb{H}$ and in which the two seemingly different types of hyperbolic line are unified is the space that is obtained from $\mathbb{C}$ by adding a single point. This is the classical construction from Complex Analysis of the *Riemann sphere* $\overline{\mathbb{C}}$.

As a set of points, the Riemann sphere is the union

$$\overline{\mathbb{C}} = \mathbb{C} \cup \{\infty\}$$

of the complex plane $\mathbb{C}$ with a point not contained in $\mathbb{C}$, which we denote ∞. In order to explore the basic properties of $\overline{\mathbb{C}}$, we first define what it means for a subset of $\overline{\mathbb{C}}$ to be open.

We begin by recalling that a set X in $\mathbb{C}$ is *open* if for each $z \in X$, there exists some $\varepsilon > 0$ so that $U_\varepsilon(z) \subset X$, where

$$U_\varepsilon(z) = \{w \in \mathbb{C} \,:\, |w - z| < \varepsilon\}$$

is the Euclidean disc of radius ε centred at z.

A set X in $\mathbb{C}$ is *closed* if its complement $\mathbb{C} - X$ in $\mathbb{C}$ is open.

A set X in $\mathbb{C}$ is *bounded* if there exists some constant $\varepsilon > 0$ so that $X \subset U_\varepsilon(0)$.

Exercise 1.8

Prove that $\mathbb{H}$ is open in $\mathbb{C}$. For each point z of $\mathbb{H}$, calculate the maximum ε so that $U_\varepsilon(z)$ is contained in $\mathbb{H}$.

In order to extend this definition to $\overline{\mathbb{C}}$, we need only define what $U_\varepsilon(z)$ means for each point z of $\overline{\mathbb{C}}$ and each $\varepsilon > 0$. Since all but one point of $\overline{\mathbb{C}}$ lies in $\mathbb{C}$, it makes sense to use the definition we had above wherever possible, and so for each point z of $\mathbb{C}$ we define

$$U_\varepsilon(z) = \{w \in \mathbb{C} \,:\, |w - z| < \varepsilon\}.$$

It remains only to define $U_\varepsilon(\infty)$, which we take to be

$$U_\varepsilon(\infty) = \{w \in \mathbb{C} \,:\, |w| > \varepsilon\} \cup \{\infty\}.$$

Definition 1.5

Say that a set X in $\overline{\mathbb{C}}$ is *open* if for each point x of X, there exists some $\varepsilon > 0$ (which may depend on x) so that $U_\varepsilon(x) \subset X$.

One immediate consequence of this definition of an open set in $\overline{\mathbb{C}}$ is that if D is an open set in $\mathbb{C}$, then D is also open in $\overline{\mathbb{C}}$. That is, we are not distorting $\mathbb{C}$ by viewing it as a subset of $\overline{\mathbb{C}}$. For example, since $\mathbb{H}$ is an open subset of $\mathbb{C}$, by Exercise 1.8, we immediately have that $\mathbb{H}$ is open in $\overline{\mathbb{C}}$ as well.

As another example, we show that the set $E = \{z \in \mathbb{C} \,:\, |z| > 1\} \cup \{\infty\}$ is open in $\overline{\mathbb{C}}$. We need to show that for each point z of E, there is some $\varepsilon > 0$ so that $U_\varepsilon(z) \subset E$.

Since $E = U_1(\infty)$, we can find a suitable ε for $z = \infty$, namely $\varepsilon = 1$. For a point z of $E - \{\infty\}$, note that the Euclidean distance from z to $\partial E = \mathbb{S}^1$ is $|z| - 1$, and so we have that $U_\varepsilon(z) \subset E$ for any $0 < \varepsilon < |z| - 1$.

On the other hand, the unit circle $\mathbb{S}^1$ in $\mathbb{C}$ is not open. No matter which point z of $\mathbb{S}^1$ and which $\varepsilon > 0$ we consider, we have that $U_\varepsilon(z)$ does not lie in $\mathbb{S}^1$, as $U_\varepsilon(z)$ necessarily contains the point $(1+\frac{1}{2}\varepsilon)z$ whose modulus is $|(1+\frac{1}{2}\varepsilon)z| = (1+\frac{1}{2}\varepsilon)|z| = 1+\frac{1}{2}\varepsilon > 1$.

Definition 1.6

A set X in $\overline{\mathbb{C}}$ is *closed* if its complement $\overline{\mathbb{C}} - X$ in $\overline{\mathbb{C}}$ is open.

For example, the unit circle $\mathbb{S}^1$ is closed in $\overline{\mathbb{C}}$, since its complement is the union

$$\overline{\mathbb{C}} - \mathbb{S}^1 = U_1(0) \cup U_1(\infty).$$

Exercise 1.9

Prove that if K is a closed and bounded subset of $\mathbb{C}$, then $X = (\mathbb{C} - K) \cup \{\infty\}$ is open in $\overline{\mathbb{C}}$. Conversely, prove that every open subset of $\overline{\mathbb{C}}$ is either an open subset of $\mathbb{C}$ or is the complement in $\overline{\mathbb{C}}$ of a closed and bounded subset of $\mathbb{C}$.

One major use of open sets is to define *convergence*. Convergence in $\overline{\mathbb{C}}$ is analogous to convergence in $\mathbb{C}$; that is, a sequence $\{z_n\}$ of points in $\overline{\mathbb{C}}$ *converges* to a point z of $\overline{\mathbb{C}}$ if for each $\varepsilon > 0$, there exists N so that $z_n \in U_\varepsilon(z)$ for all $n > N$.

Exercise 1.10

Prove that $\{z_n = \frac{1}{n} \,|\, n \in \mathbb{N}\}$ converges to 0 in $\overline{\mathbb{C}}$, and that $\{w_n = n \,|\, n \in \mathbb{N}\}$ converges to ∞ in $\overline{\mathbb{C}}$.

Let X be a subset of $\overline{\mathbb{C}}$. Define the *closure* $\overline{X}$ of X in $\overline{\mathbb{C}}$ to be the set

$$\overline{X} = \{z \in \overline{\mathbb{C}} \,|\, U_\varepsilon(z) \cap X \neq \emptyset \text{ for all } \varepsilon > 0\}.$$

Note that every point $x \in X$ lies in $\overline{X}$, since $\{x\} \subset U_\varepsilon(x) \cap X$ for every $\varepsilon > 0$. There may be points in $\overline{X}$ other than the points of X.

In particular, note that if $\{x_n\}$ is a sequence of points of X converging to a point x of $\overline{\mathbb{C}}$, then x is necessarily a point of $\overline{X}$.

Exercise 1.11

Determine the closure in $\overline{\mathbb{C}}$ of $X = \{\frac{1}{n} \,|\, n \in \mathbb{Z} - \{0\}\}$ and of $Y = \mathbb{Q} + \mathbb{Q}i = \{a + b\,i \,|\, a, b \in \mathbb{Q}\}$.

Exercise 1.12

If X is a subset of $\overline{\mathbb{C}}$, prove that $\overline{X}$ is closed in $\overline{\mathbb{C}}$.

We are now ready to unify the two notions of Euclidean line and Euclidean circle in $\mathbb{C}$.

Definition 1.7

A *circle in* $\overline{\mathbb{C}}$ is either a Euclidean circle in $\mathbb{C}$, or the union of a Euclidean line in $\mathbb{C}$ with $\{\infty\}$.

That is, we use the point ∞, which we adjoined to $\mathbb{C}$ to obtain $\overline{\mathbb{C}}$, to be the point we add to each Euclidean line to get a circle.

As a bit of notation, for a Euclidean line L in $\mathbb{C}$, let $\overline{L} = L \cup \{\infty\}$ be the circle in $\overline{\mathbb{C}}$ containing L. For example, the *extended real axis* $\overline{\mathbb{R}} = \mathbb{R} \cup \{\infty\}$ is the circle in $\overline{\mathbb{C}}$ containing the real axis $\mathbb{R}$ in $\mathbb{C}$.

Note that this notation for the circle in $\overline{\mathbb{C}}$ containing the Euclidean line L agrees with our earlier notation for the closure of a subset of $\overline{\mathbb{C}}$, as the closure in $\overline{\mathbb{C}}$ of a Euclidean line L in $\mathbb{C}$ is exactly $L \cup \{\infty\}$.

As might be guessed, there is a generalization of stereographic projection to the Riemann sphere and the complex plane.

Identify $\mathbb{C}$ with the x_1x_2 plane in $\mathbb{R}^3$, where the coordinates on $\mathbb{R}^3$ are (x_1, x_2, x_3), by identifying the point $z = x + iy$ in $\mathbb{C}$ with the point $(x, y, 0)$ in $\mathbb{R}^3$. Let $\mathbb{S}^2$ be the unit sphere in $\mathbb{R}^3$, that is

$$\mathbb{S}^2 = \{(x, y, z) \in \mathbb{R}^3 \mid x^2 + y^2 + z^2 = 1\},$$

with north pole $\mathrm{N} = (0, 0, 1)$.

Consider the function $\xi : \mathbb{S}^2 - \{\mathrm{N}\} \to \mathbb{C}$ defined as follows. For each point P of $\mathbb{S}^2 - \{\mathrm{N}\}$, let L_{P} be the Euclidean line in $\mathbb{R}^3$ passing through N and P, and define $\xi(\mathrm{P})$ to be the point of intersection $L_{\mathrm{P}} \cap \mathbb{C}$.

Exercise 1.13

Write out explicit formulae for both ξ and its inverse $\xi^{-1} : \mathbb{C} \to \mathbb{S}^2 - \{\mathrm{N}\}$.

The bijectivity of ξ follows from the fact that we are able to write down an explicit expression for ξ^{-1}. We could also argue geometrically, as we did for stereographic projection from $\mathbb{S}^1 - \{i\}$ to $\mathbb{R}$.

We are also able to describe circles in $\overline{\mathbb{C}}$ as the sets of solutions to equations in $\overline{\mathbb{C}}$. Recall that we show in Exercise 1.1 that every Euclidean circle in $\mathbb{C}$ can be described as the set of solutions of an equation of the form

$$\alpha z\overline{z} + \beta z + \overline{\beta}\overline{z} + \gamma = 0,$$

where α, $\gamma \in \mathbb{R}$ and $\beta \in \mathbb{C}$, and that every Euclidean line in $\mathbb{C}$ can be described as the set of solutions of an equation of the form

$$\beta z + \overline{\beta}\overline{z} + \gamma = 0,$$

where $\gamma \in \mathbb{R}$ and $\beta \in \mathbb{C}$.

Combining these, we see that every circle in $\overline{\mathbb{C}}$ can be described as the set of solutions in $\overline{\mathbb{C}}$ to an equation of the form

$$\alpha z\overline{z} + \beta z + \overline{\beta}\overline{z} + \gamma = 0,$$

where α, $\gamma \in \mathbb{R}$ and $\beta \in \mathbb{C}$.

There is one subtlety to be considered here, namely the question of how we consider whether ∞ is or is not a solution of such an equation.

For an equation of the form

$$\beta z + \overline{\beta}\overline{z} + \gamma = 0,$$

we may consider ∞ to be a solution *by continuity*. That is, there is a sequence $\{z_n\}$ of points in $\mathbb{C}$ that satisfies this equation and that converges to ∞ in $\overline{\mathbb{C}}$.

Specifically, let w_0 and w_1 be two distinct solutions, so that every linear combination of the form $w_0 + t(w_1 - w_0)$, $t \in \mathbb{R}$, is also a solution. Consider the sequence

$$\{z_n = w_0 + n(w_1 - w_0),\ n \in \mathbb{N}\}.$$

This sequence converges to ∞ in $\overline{\mathbb{C}}$, and for each n we have that

$$\beta z_n + \overline{\beta}\overline{z_n} + \gamma = 0.$$

However, for an equation of the form

$$\alpha z\overline{z} + \beta z + \overline{\beta}\overline{z} + \gamma = 0,\ \alpha \neq 0,$$

we cannot view ∞ as a solution to the equation by continuity. This follows immediately from the fact that we can rewrite

$$\alpha z\overline{z} + \beta z + \overline{\beta}\overline{z} + \gamma = \alpha \left| z + \frac{\overline{\beta}}{\alpha} \right|^2 + \gamma - \frac{|\beta|^2}{\alpha}.$$

In particular, if $\{z_n\}$ is any sequence of points in $\overline{\mathbb{C}}$ converging to ∞, then

$$\lim_{n\to\infty} (\alpha z_n\overline{z_n} + \beta z_n + \overline{\beta}\overline{z_n} + \gamma) = \infty.$$

Therefore, z_n cannot lie on the circle

$$A = \{z \in \mathbb{C} \mid \alpha z\overline{z} + \beta z + \overline{\beta}\overline{z} + \gamma = 0\}$$

for n large, and so we cannot consider ∞ to be a point of A.

Since we now have a definition of what it means for a subset of $\overline{\mathbb{C}}$ to be open, we are able to define what it means for a function $f : \overline{\mathbb{C}} \to \overline{\mathbb{C}}$ to be continuous, this time in analogy with the usual definition of continuity of functions from $\mathbb{R}$ to $\mathbb{R}$.

Definition 1.8

A function $f : \overline{\mathbb{C}} \to \overline{\mathbb{C}}$ is *continuous at* $z \in \overline{\mathbb{C}}$ if for each $\varepsilon > 0$, there exists $\delta > 0$ so that $w \in U_\delta(z)$ implies that $f(w) \in U_\varepsilon(f(z))$. A function $f : \overline{\mathbb{C}} \to \overline{\mathbb{C}}$ is *continuous* if it is continuous at every point z of $\overline{\mathbb{C}}$.

One advantage to generalizing this definition of continuity is that we may use exactly the same proofs as with functions from $\mathbb{R}$ to $\mathbb{R}$ to show that constant functions from $\overline{\mathbb{C}}$ to $\overline{\mathbb{C}}$ are continuous, as are products and quotients (when they are defined), sums and differences (when they are defined), and compositions of continuous functions.

However, there are some slight differences between functions from $\mathbb{R}$ to $\mathbb{R}$ and functions from $\overline{\mathbb{C}}$ to $\overline{\mathbb{C}}$, which arise from the presence of the point ∞. Consider the following example.

Proposition 1.9

The function $J : \overline{\mathbb{C}} \to \overline{\mathbb{C}}$ defined by

$$J(z) = \frac{1}{z} \text{ for } z \in \mathbb{C} - \{0\}, \quad J(0) = \infty, \text{ and } J(\infty) = 0$$

is continuous on $\overline{\mathbb{C}}$.

To see that J is continuous at 0, take $\varepsilon > 0$ to be given. Since we have that $J(0) = \infty$, we need to show that there exists some $\delta > 0$ so that

$$J(U_\delta(0)) \subset U_\varepsilon(J(0)) = U_\varepsilon(\infty).$$

Take $\delta = \frac{1}{\varepsilon}$. For each $w \in U_\delta(0) - \{0\}$, we have that

$$|J(w)| = \frac{1}{|w|} > \frac{1}{\delta} = \varepsilon,$$

and so $J(w) \in U_\varepsilon(\infty)$. Since we have that $J(0) = \infty \in U_\varepsilon(\infty)$ by definition, we see that J is continuous at 0.

The argument that J is continuous at ∞ is very similar to the argument that J is continuous at 0. Again, given $\varepsilon > 0$ we take $\delta = \frac{1}{\varepsilon}$. Then, for each $w \in U_\delta(\infty) - \{\infty\}$, we have that

$$|J(w)| = \frac{1}{|w|} < \frac{1}{\delta} = \varepsilon,$$

and so $J(w) \in U_\varepsilon(0)$. We have that $J(\infty) = 0 \in U_\varepsilon(0)$ by definition, and so J is continuous at ∞.

To complete the proof, let $z \in \mathbb{C} - \{0\}$ be any point, and let $\varepsilon > 0$ be given. We need to find $\delta > 0$ so that $w \in U_\delta(z)$ implies that $J(w) \in U_\varepsilon(J(z))$. Let $\varepsilon' = \min(\varepsilon, \frac{1}{2|z|})$, so that $U_{\varepsilon'}(z)$ does not contain 0.

For any $\xi \in U_{\varepsilon'}(J(z))$, we have that

$$|\xi| < |J(z)| + \varepsilon' = \frac{1}{|z|} + \varepsilon'.$$

Since $\varepsilon' \leq \frac{1}{2|z|}$, we have that

$$|\xi| < \frac{3}{2|z|}.$$

Writing $\xi = \frac{1}{w}$, this gives that

$$\frac{1}{|w|} < \frac{3}{2|z|}, \text{ and so } \frac{1}{|zw|} < \frac{3}{2|z|^2}.$$

So, set $\delta = \frac{2}{3}\varepsilon'|z|^2$.

For $|z - w| < \delta$, we then have that

$$|J(z) - J(w)| = \left|\frac{1}{z} - \frac{1}{w}\right| = \frac{|z-w|}{|zw|} < \frac{2}{3}\varepsilon'|z|^2\frac{3}{2|z|^2} = \varepsilon'.$$

Since $\varepsilon' \leq \varepsilon$, we have that J is continuous at $z \in \mathbb{C} - \{0\}$. This completes the proof of Proposition 1.9.

Exercise 1.14

Let $g(z)$ be a polynomial. Prove that the function $f : \overline{\mathbb{C}} \to \overline{\mathbb{C}}$, defined by

$$f(z) = g(z) \text{ for } z \in \mathbb{C} \text{ and } f(\infty) = \infty,$$

is continuous on $\overline{\mathbb{C}}$.

One very useful property, and indeed a defining property, of continuous functions is that they preserve convergent sequences. That is, if $f : \overline{\mathbb{C}} \to \overline{\mathbb{C}}$ is a continuous function and if $\{x_n\}$ is a sequence in $\overline{\mathbb{C}}$ converging to x, then $\{f(x_n)\}$ converges to $f(x)$.

There is a class of continuous functions from $\overline{\mathbb{C}}$ to itself that are especially well behaved.

Definition 1.10

A function $f : \overline{\mathbb{C}} \to \overline{\mathbb{C}}$ is a *homeomorphism* if f is a bijection and if both f and f^{-1} are continuous.

We have already seen one example of a homeomorphism of $\overline{\mathbb{C}}$.

Proposition 1.11

The function $J : \overline{\mathbb{C}} \to \overline{\mathbb{C}}$ defined by

$$J(z) = \frac{1}{z} \text{ for } z \in \mathbb{C} - \{0\}, \quad J(0) = \infty, \text{ and } J(\infty) = 0,$$

is a homeomorphism of $\overline{\mathbb{C}}$.

Since $J \circ J(z) = z$ for all $z \in \overline{\mathbb{C}}$, we immediately have that J is bijective. To see that J is injective, suppose that there exist points z and w for which $J(z) = J(w)$, and note that $z = J(J(z)) = J(J(w)) = w$. To see that J is surjective, note that for any $z \in \overline{\mathbb{C}}$, we have that $z = J(J(z))$.

Moreover, since $J^{-1}(z) = J(z)$ for all $z \in \overline{\mathbb{C}}$ and since J is continuous, by Proposition 1.9, we see that J^{-1} is continuous. This completes the proof of Proposition 1.11.

The homeomorphisms of $\overline{\mathbb{C}}$ are the transformations of $\overline{\mathbb{C}}$ that are of most interest to us, so set

$$\text{Homeo}(\overline{\mathbb{C}}) = \{f : \overline{\mathbb{C}} \to \overline{\mathbb{C}} \mid f \text{ is a homeomorphism}\}.$$

By definition, the inverse of a homeomorphism is again a homeomorphism. Also, the composition of two homeomorphisms is again a homeomorphism, since the composition of bijections is again a bijection and since the composition of continuous functions is again continuous. As the identity homeomorphism $f : \overline{\mathbb{C}} \to \overline{\mathbb{C}}$ given by $f(z) = z$ is a homeomorphism, we have that $\text{Homeo}(\overline{\mathbb{C}})$ is a group.

Exercise 1.15

Let $g(z)$ be a polynomial. Prove that the function $f : \overline{\mathbb{C}} \to \overline{\mathbb{C}}$, defined by

$$f(z) = g(z) \text{ for } z \in \mathbb{C} \text{ and } f(\infty) = \infty,$$

is a homeomorphism if and only if the degree of g is one.

Exercise 1.16

A subset X of $\overline{\mathbb{C}}$ is *dense* if $\overline{X} = \overline{\mathbb{C}}$. Prove that if X is dense in $\overline{\mathbb{C}}$ and if $f : \overline{\mathbb{C}} \to \overline{\mathbb{C}}$ is a continuous function so that $f(x) = x$ for all x in X, then $f(z) = z$ for all z in $\overline{\mathbb{C}}$.

1.3 The Boundary at Infinity of $\mathbb{H}$

In Section 1.2, we define a circle in the Riemann sphere $\overline{\mathbb{C}}$ to be either a Euclidean circle in $\mathbb{C}$ or the union of a Euclidean line in $\mathbb{C}$ with $\{\infty\}$. We also have several examples of circles in $\overline{\mathbb{C}}$, including the *unit circle* $\mathbb{S}^1$ in $\mathbb{C}$ and the *extended real axis* $\overline{\mathbb{R}} = \mathbb{R} \cup \{\infty\}$.

In particular, the complement of a circle in $\overline{\mathbb{C}}$ has two components. For $\mathbb{S}^1$, the components of $\overline{\mathbb{C}} - \mathbb{S}^1$ are the Euclidean disc $\mathbb{D} = U_1(0)$ and the disc $U_1(\infty)$, while for $\overline{\mathbb{R}}$ the components of $\overline{\mathbb{C}} - \overline{\mathbb{R}}$ are the upper half-plane $\mathbb{H}$ and the lower half-plane $\{z \in \mathbb{C} \mid \text{Im}(z) < 0\}$.

Definition 1.12

Define a *disc in* $\overline{\mathbb{C}}$ to be one of the components of the complement in $\overline{\mathbb{C}}$ of a circle in $\overline{\mathbb{C}}$.

Note that every disc in $\overline{\mathbb{C}}$ determines a unique circle in $\overline{\mathbb{C}}$, and that every circle in $\overline{\mathbb{C}}$ determines two disjoint discs in $\overline{\mathbb{C}}$.

For the remainder of this section, we focus our attention on one particular disc in $\overline{\mathbb{C}}$, namely $\mathbb{H}$, and the circle in $\overline{\mathbb{C}}$ determining it, namely $\overline{\mathbb{R}}$. We refer to $\overline{\mathbb{R}}$ as the *boundary at infinity* of $\mathbb{H}$, and we refer to points of $\overline{\mathbb{R}}$ as *points at infinity* of $\mathbb{H}$. The reason for using this term will be explained in Section 3.7, after we have developed a means of measuring distance in $\mathbb{H}$.

More generally, for any set X in $\mathbb{H}$, we can make sense of the notion of the *boundary at infinity* of X. Specifically, we form the closure $\overline{X}$ of X in $\overline{\mathbb{C}}$, and then define the *boundary at infinity of* X to be the intersection $\overline{X} \cap \overline{\mathbb{R}}$ of $\overline{X}$ with the boundary at infinity $\overline{\mathbb{R}}$ of $\mathbb{H}$.

As an example, let ℓ be a hyperbolic line in $\mathbb{H}$, and suppose that ℓ is contained in the circle A in $\overline{\mathbb{C}}$. Then, the boundary at infinity of ℓ is the pair of points contained in the intersection $A \cap \overline{\mathbb{R}}$.

There are more complicated examples as well. Let ℓ_1 and ℓ_2 be parallel hyperbolic lines in $\mathbb{H}$, and let H be the region in $\mathbb{H}$ which consists of the two lines ℓ_1 and ℓ_2, together with the part of $\mathbb{H}$ which lies between them. There are two possibilities for the boundary at infinity of this region H.

Let C_k be the circle in $\overline{\mathbb{C}}$ containing ℓ_k. Since ℓ_1 and ℓ_2 are disjoint, either C_1 and C_2 are disjoint, or C_1 and C_2 intersect in a single point, which is then necessarily contained in $\overline{\mathbb{R}}$.

In the case that C_1 and C_2 intersect at the point x of $\overline{\mathbb{R}}$, the boundary at infinity of H is the union of a closed arc in $\overline{\mathbb{R}}$ and the set $\{x\}$.

In the case that C_1 and C_2 are disjoint, the boundary at infinity of H is the union of two closed arcs in $\overline{\mathbb{R}}$. These two possibilities are shown in Fig. 1.5.

Figure 1.5: Two possibilities for parallel hyperbolic lines

This gives us a way of distinguishing two different types of parallelism for hyperbolic lines in $\mathbb{H}$. Namely, there are parallel hyperbolic lines whose boundaries at infinity intersect, and there are parallel hyperbolic lines whose boundaries at infinity are disjoint. When we need to make the distinction, we refer to a pair of the latter type as *ultraparallel.*

There is another way to see the distinction between parallel and ultraparallel hyperbolic lines.

Exercise 1.17

Let ℓ_1 and ℓ_2 be parallel hyperbolic lines. Show that ℓ_1 and ℓ_2 are ultraparallel if and only if there exists a hyperbolic line perpendicular to both ℓ_1 and ℓ_2.

We saw in Section 1.1, specifically in Proposition 1.2, that two points in $\mathbb{H}$ determine a unique hyperbolic line in $\mathbb{H}$. The key to the proof of this fact is that there is a unique Euclidean circle or Euclidean line in $\mathbb{C}$ that passes through the given two points and that is perpendicular to the real axis $\mathbb{R}$.

This same argument applies to hyperbolic lines determined by points at infinity.

Proposition 1.13

Let p be a point of $\mathbb{H}$ and q a point of $\overline{\mathbb{R}}$. Then, there is a unique hyperbolic line in $\mathbb{H}$ determined by p and q.

Suppose that $q = \infty$. Of all the hyperbolic lines through p, there is exactly one that contains q in its boundary at infinity, namely the hyperbolic line contained in the Euclidean line $\{z \in \mathbb{C} | \text{Re}(z) = \text{Re}(p)\}$. The statement about uniqueness follows from the observation that no hyperbolic line contained in a Euclidean circle contains ∞ in its boundary at infinity.

Suppose that $q \neq \infty$ and that $\text{Re}(p) = \text{Re}(q)$. Then, the hyperbolic line contained in the Euclidean line $\{z \in \mathbb{C} \mid \text{Re}(z) = \text{Re}(p)\}$ is the unique hyperbolic line through p that contains q in its boundary at infinity.

Suppose that $q \neq \infty$ and that $\text{Re}(p) \neq \text{Re}(q)$. Then, we may again use the construction from the proof of Proposition 1.2 of the perpendicular bisector of the Euclidean line segment joining p to q to find the unique Euclidean circle centred on the real axis $\mathbb{R}$ that passes through both p and q. Intersecting this circle with $\mathbb{H}$ yields the unique hyperbolic line determined by p and q. This completes the proof of Proposition 1.13.

We refer to the part of the hyperbolic line between p and q as the *hyperbolic ray determined by* p *and* q , or as the *hyperbolic ray through* p *with endpoint at infinity* q.

The argument for the existence and uniqueness of a hyperbolic line determined by two points at infinity is similar, and is left as an exercise.

Exercise 1.18

Let p and q be two points of $\overline{\mathbb{R}}$. Prove that p and q determine a unique hyperbolic line whose endpoints at infinity are p and q.

2
The General Möbius Group

As our goal is to study the geometry of the hyperbolic plane by considering quantities invariant under the action of a reasonable group of transformations, we spend this chapter by describing such a reasonable group of transformations of $\overline{\mathbb{C}}$, namely the *general Möbius group* Möb, which consists of compositions of *Möbius transformations* and *reflections*. We close the chapter by restricting our attention to the transformations in Möb preserving $\mathbb{H}$.

2.1 The Group of Möbius Transformations

Since every hyperbolic line in $\mathbb{H}$ is by definition contained in a circle in $\overline{\mathbb{C}}$, we begin the process of determining the transformations of $\mathbb{H}$ taking hyperbolic lines to hyperbolic lines by first determining the group of homeomorphisms of $\overline{\mathbb{C}}$ taking circles in $\overline{\mathbb{C}}$ to circles in $\overline{\mathbb{C}}$.

For the sake of notational convenience, let $\text{Homeo}^{\text{C}}(\overline{\mathbb{C}})$ be the subset of the group $\text{Homeo}(\overline{\mathbb{C}})$ of homeomorphisms of $\overline{\mathbb{C}}$ that contains all those homeomorphisms of $\overline{\mathbb{C}}$ taking circles in $\overline{\mathbb{C}}$ to circles in $\overline{\mathbb{C}}$.

Note that, while it is easy to see that the composition of two elements of $\text{Homeo}^{\text{C}}(\overline{\mathbb{C}})$ is again an element of $\text{Homeo}^{\text{C}}(\overline{\mathbb{C}})$ and that the identity homeomorphism is an element of $\text{Homeo}^{\text{C}}(\overline{\mathbb{C}})$, we do not yet know that inverses of elements of $\mathrm{Homeo}^{\mathrm{C}}(\overline{\mathbb{C}})$ lie in $\mathrm{Homeo}^{\mathrm{C}}(\overline{\mathbb{C}})$, and hence we cannot yet conclude that $\mathrm{Homeo}^{\mathrm{C}}(\overline{\mathbb{C}})$ is a group.

In fact, there are many homeomorphisms of $\overline{\mathbb{C}}$ that do not lie in $\mathrm{Homeo}^{\mathrm{C}}(\overline{\mathbb{C}})$.

Exercise 2.1

Give an explicit example of an element of $\mathrm{Homeo}(\overline{\mathbb{C}})$ that is not an element of $\mathrm{Homeo}^{\mathrm{C}}(\overline{\mathbb{C}})$.

We begin by considering a class of homeomorphisms of $\overline{\mathbb{C}}$ that we understand, namely those arising from polynomials. As we saw in Exercise 1.14 and Exercise 1.15, to each polynomial $g(z)$ we may associate the function $f : \overline{\mathbb{C}} \to \overline{\mathbb{C}}$ given by

$$f(z) = g(z) \text{ for } z \in \mathbb{C} \text{ and } f(\infty) = \infty.$$

As we wish to consider homeomorphisms of $\overline{\mathbb{C}}$ which arise from polynomials, we restrict our attention to polynomials of degree 1.

Proposition 2.1

The element f of $\mathrm{Homeo}(\overline{\mathbb{C}})$ defined by

$$f(z) = az + b \text{ for } z \in \mathbb{C} \text{ and } f(\infty) = \infty,$$

where a, $b \in \mathbb{C}$ and $a \neq 0$, is an element of $\mathrm{Homeo}^{\mathrm{C}}(\overline{\mathbb{C}})$.

Recall from Section 1.2 that each circle A in $\overline{\mathbb{C}}$ is the set of solutions to an equation of the form

$$\alpha z\overline{z} + \beta z + \overline{\beta}\overline{z} + \gamma = 0,$$

where α, $\gamma \in \mathbb{R}$ and $\beta \in \mathbb{C}$, and where $\alpha \neq 0$ if and only if A is a circle in $\mathbb{C}$.

We begin with the case that A is a Euclidean line in $\mathbb{C}$. So, consider the Euclidean line A given as the solution to the equation

$$A = \{z \in \mathbb{C} \mid \beta z + \overline{\beta}\overline{z} + \gamma = 0\},$$

where $\beta \in \mathbb{C}$ and $\gamma \in \mathbb{R}$. We wish to show that if z satisfies this equation, then $w = az + b$ satisfies a similar equation.

Since $w = az+b$, we have that $z = \frac{1}{a}(w-b)$. Substituting this into the equation for A given above gives

$$\begin{aligned}\beta z + \overline{\beta}\overline{z} + \gamma &= \beta\frac{1}{a}(w-b) + \overline{\beta}\overline{\frac{1}{a}(w-b)} + \gamma \\ &= \frac{\beta}{a}w + \overline{\left(\frac{\beta}{a}\right)}\overline{w} - \frac{\beta}{a}b - \overline{\frac{\beta}{a}b} + \gamma = 0.\end{aligned}$$

Since $\frac{\beta}{a}b+\overline{\frac{\beta}{a}b}=2\,\mathrm{Re}\left(\frac{\beta}{a}b\right)$ is real, this shows that w also satisfies the equation of a Euclidean line. Hence, f takes Euclidean lines in $\mathbb{C}$ to Euclidean lines in $\mathbb{C}$. The proof that f takes Euclidean circles to Euclidean circles is similar, and is left as an exercise.

Exercise 2.2

Show that the homeomorphism $f:\overline{\mathbb{C}}\to\overline{\mathbb{C}}$ defined by setting

$$f(z)=az+b \text{ for } z\in\mathbb{C} \text{ and } f(\infty)=\infty,$$

where a, $b\in\mathbb{C}$ and $a\neq 0$, takes Euclidean circles in $\mathbb{C}$ to Euclidean circles in $\mathbb{C}$.

Exercise 2.2 completes the proof of Proposition 2.1.

We can refine this argument to obtain quantitative information about the image circle in $\overline{\mathbb{C}}$ in terms of the coefficients of $f(z)=az+b$ and the equation of the original circle in $\overline{\mathbb{C}}$.

For example, suppose that L is a Euclidean line given by the equation $\beta z+\overline{\beta}\overline{z}+\gamma=0$, and recall from the solution of Exercise 1.1 that the slope of L is $\frac{\mathrm{Re}(\beta)}{\mathrm{Im}(\beta)}$.

We have seen that f takes L to the Euclidean line $f(L)$ given by the equation

$$\frac{\beta}{a}w+\overline{\left(\frac{\beta}{a}\right)}\overline{w}-\frac{\beta}{a}b-\overline{\frac{\beta}{a}}\overline{b}+\gamma=0,$$

which has slope $\frac{\mathrm{Re}(\beta\overline{a})}{\mathrm{Im}(\beta\overline{a})}$.

Exercise 2.3

Determine the Euclidean centre and Euclidean radius of the image of the Euclidean circle A given by the equation $\alpha z\overline{z}+\beta z+\overline{\beta}\overline{z}+\gamma=0$ under the homeomorphism

$$f(z)=az+b \text{ for } z\in\mathbb{C} \text{ and } f(\infty)=\infty,$$

where a, $b\in\mathbb{C}$ and $a\neq 0$.

There is another homeomorphism of $\overline{\mathbb{C}}$ we considered earlier in Proposition 1.11, namely the function $J:\overline{\mathbb{C}}\to\overline{\mathbb{C}}$ defined by setting

$$J(z)=\frac{1}{z} \text{ for } z\in\mathbb{C}-\{0\},\ \ J(0)=\infty, \text{ and } J(\infty)=0.$$

Proposition 2.2

The element J of $\text{Homeo}(\overline{\mathbb{C}})$ defined by

$$J(z) = \frac{1}{z} \text{ for } z \in \mathbb{C} - \{0\}, \ J(0) = \infty, \text{ and } J(\infty) = 0,$$

is an element of $\text{Homeo}^{\mathrm{C}}(\overline{\mathbb{C}})$.

We proceed as before. Let A be a circle in $\overline{\mathbb{C}}$ given by the equation $\alpha z\overline{z} + \beta z + \overline{\beta}\overline{z} + \gamma = 0$, where $\alpha, \gamma \in \mathbb{R}$ and $\beta \in \mathbb{C}$.

Set $w = \frac{1}{z}$, so that $z = \frac{1}{w}$. Substituting this back into the equation for A gives

$$\alpha \frac{1}{w}\overline{\frac{1}{w}} + \beta \frac{1}{w} + \overline{\beta}\overline{\frac{1}{w}} + \gamma = 0.$$

Multiplying through by $w\overline{w}$ we see that w satisfies the equation

$$\alpha + \beta\overline{w} + \overline{\beta}w + \gamma w\overline{w} = 0.$$

Since α and γ are real and since the coefficients of w and $\overline{w}$ are conjugate, this is again the equation of a circle in $\overline{\mathbb{C}}$. This completes the proof of Proposition 2.2.

As in the proof of Proposition 2.1, we can extract some quantitiative information from the proof of Proposition 2.2 about the circle $J(A)$ in terms of the circle A.

For example, if A is the circle in $\overline{\mathbb{C}}$ given by the equation $2z + 2\overline{z} + 3 = 0$, then $J(A)$ is the circle in $\overline{\mathbb{C}}$ given by the equation $2\overline{w} + 2w + 3w\overline{w} = 0$, which is a Euclidean circle in $\mathbb{C}$ with Euclidean centre $-\frac{2}{3}$ and Euclidean radius $\frac{2}{3}$.

Exercise 2.4

Let A be a Euclidean circle in $\mathbb{C}$ given by the equation $|z - z_0| = r$. Determine conditions on z_0 and r so that $J(A)$ is a Euclidean line in $\mathbb{C}$.

Note that every possible composition of these two types of homeomorphisms of $\overline{\mathbb{C}}$, namely the $f(z) = az + b$ with $a, b \in \mathbb{C}$ and $a \neq 0$, and $J(z) = \frac{1}{z}$, have the form $m(z) = \frac{az+b}{cz+d}$. This leads us to the following definition.

Definition 2.3

A *Möbius transformation* is a function $m : \overline{\mathbb{C}} \to \overline{\mathbb{C}}$ of the form

$$m(z) = \frac{az + b}{cz + d},$$

where a, b, c, $d \in \mathbb{C}$ and $ad - bc \neq 0$. Let Möb^+ denote the set of all Möbius transformations.

We pause here to insert a remark about the arithmetic of ∞. For any $a \neq 0$, we can unambiguously assign the value of $\frac{a}{0}$ to be ∞ by continuity. That is, we set

$$\frac{a}{0} = \lim_{w \to 0} \frac{a}{w}.$$

Since $a \neq 0$, $\frac{a}{w}$ is non-zero, and by considering the modulus $|\frac{a}{w}|$, we can see that $\lim_{w\to 0} \frac{a}{w} = \infty$ in $\overline{\mathbb{C}}$. However, we are still unable to make sense of the expression $\frac{0}{0}$.

Similarly, we define the image of ∞ under $m(z) = \frac{az+b}{cz+d}$ by continuity. That is, we set

$$m(\infty) = \lim_{z\to\infty} \frac{az+b}{cz+d} = \lim_{z\to\infty} \frac{a+\frac{b}{z}}{c+\frac{d}{z}} = \frac{a}{c}.$$

The value $m(\infty)$ is well-defined since one of a or c has to be non-zero, since from the definition of Möbius transformation we know that $ad - bc \neq 0$.

Observe that, since $m(\infty) = \frac{a}{c}$, we have that $m(\infty) = \infty$ if and only if $c = 0$. Further, since $m(0) = \frac{b}{d}$, we have that $m(0) = 0$ if and only if $b = 0$.

As we see in Exercise 2.5, we can write down an explicit expression for the inverse of a Möbius transformation. Since the composition of two Möbius transformations is again a Möbius transformation, we have that the set Möb^+ of Möbius transformations is a group under composition with identity element $m(z) = z$.

Exercise 2.5

In order to prove that Möbius transformations are bijective, give an explicit expression for the inverse of the Möbius transformation $m(z) = \frac{az+b}{cz+d}$.

As we have already mentioned, the form of a Möbius transformation is very similar to the forms of the homeomorphisms of $\overline{\mathbb{C}}$ we encountered earlier this section, namely

$$f(z) = az + b \text{ for } z \in \mathbb{C} \text{ and } f(\infty) = \infty,$$

where a, $b \in \mathbb{C}$ and $a \neq 0$, and also

$$J(z) = \frac{1}{z} \text{ for } z \in \mathbb{C} - \{0\},\ \ J(0) = \infty, \text{ and } J(\infty) = 0.$$

In fact, we may write any Möbius transformation $m(z) = \frac{az+b}{cz+d}$ as a composition of such homeomorphisms.

Theorem 2.4

Consider the Möbius transformation $m(z) = \frac{az+b}{cz+d}$, where $a, b, c, d \in \mathbb{C}$ and $ad - bc \neq 0$.

If $c = 0$, then $m(z) = \frac{a}{d}z + \frac{b}{d}$.

If $c \neq 0$, then $m(z) = f(J(g(z)))$, where $g(z) = c^2 z + cd$ and $f(z) = -(ad - bc)z + \frac{a}{c}$.

The proof of Theorem 2.4 is a direct calculation. If $c = 0$, there is nothing to check. If $c \neq 0$, then

$$m(z) = \frac{az+b}{cz+d} = \frac{(az+b)\,c}{(cz+d)\,c} = \frac{acz+bc}{c^2 z + cd}.$$

Since $ad - bc \neq 0$, we have that

$$m(z) = \frac{acz+bc}{c^2 z+cd} = \frac{acz + ad - (ad-bc)}{c^2 z + cd} = \frac{a}{c} - \frac{ad-bc}{c^2 z + cd} = f(J(g(z))),$$

where $g(z) = c^2 z + cd$ and $f(z) = -(ad - bc)z + \frac{a}{c}$. This completes the proof of Theorem 2.4.

Theorem 2.4 has several immediate corollaries. First, every Möbius transformation is a homeomorphism, as it is a composition of homeomorphisms. That is,

$$\text{Möb}^+ \subset \text{Homeo}(\overline{\mathbb{C}}).$$

Second, every Möbius transformation takes circles in $\overline{\mathbb{C}}$ to circles in $\overline{\mathbb{C}}$, as it is a composition of functions with this property. We combine this observation with the previous observation in the following theorem.

Theorem 2.5

$\text{Möb}^+ \subset \text{Homeo}^{\text{C}}(\overline{\mathbb{C}})$.

We note here that the condition that $ad - bc \neq 0$ in the definition of a Möbius transformation is not spurious.

Exercise 2.6

Consider a function $p : \overline{\mathbb{C}} \to \overline{\mathbb{C}}$ of the form $p(z) = \frac{az+b}{cz+d}$ where $a, b, c, d \in \mathbb{C}$ and $ad - bc = 0$. Prove that p is not a homeomorphism of $\overline{\mathbb{C}}$.

We close this section with a very crude classification of Möbius transformations, based on the number of fixed points. A *fixed point* of the Möbius transformation m is a point z of $\overline{\mathbb{C}}$ satisfying $m(z) = z$. Suppose that m is not the identity.

We saw earlier in this section that for $m(z) = \frac{az+b}{cz+d}$, we have that $m(\infty) = \frac{a}{c}$, and so $m(\infty) = \infty$ if and only if $c = 0$.

If $c = 0$, then $m(z) = \frac{a}{d}z + \frac{b}{d}$, and the fixed point of m in $\mathbb{C}$ is the solution to the equation $m(z) = \frac{a}{d}z + \frac{b}{d} = z$. If $\frac{a}{d} = 1$, then there is no solution in $\mathbb{C}$, while if $\frac{a}{d} \neq 1$, then $z = \frac{b}{d-a}$ is the unique solution in $\mathbb{C}$. In particular, if $c = 0$, then m has either one or two fixed points.

If $c \neq 0$, then $m(\infty) \neq \infty$, and so the fixed points of m are the solutions in $\mathbb{C}$ of the equation $m(z) = \frac{az+b}{cz+d} = z$, which are the roots of the quadratic polynomial $cz^2 + (d-a)z - b = 0$. In particular, if $c \neq 0$, then again m has either one or two fixed points.

This analysis has the following important consequence.

Theorem 2.6

Let $m(z)$ be a Möbius transformation fixing three distinct points of $\overline{\mathbb{C}}$. Then, m is the identity transformation. That is, $m(z) = z$ for every point z of $\overline{\mathbb{C}}$.

> *Exercise 2.7*
>
> Calculate the fixed points of each of the following Möbius transformations.
>
> 1. $m(z) = \frac{2z+5}{3z-1}$; 2. $m(z) = 7z + 6$; 3. $J(z) = \frac{1}{z}$; 4. $m(z) = \frac{z}{z+1}$.

2.2 Transitivity Properties of Möb$^+$

One of the most basic properties of Möb$^+$ is that it acts *uniquely triply transitively* on $\overline{\mathbb{C}}$. By this we mean that given two triples (z_1, z_2, z_3) and (w_1, w_2, w_3) of distinct points of $\overline{\mathbb{C}}$, there exists a unique element m of Möb$^+$ so that $m(z_1) = w_1$, $m(z_2) = w_2$, and $m(z_3) = w_3$.

As is often the case, we begin a proof of existence and uniqueness by first showing uniquness. We then construct a particular transformation by whatever means are at hand, and observe that by uniqueness it must be the only one.

So, given two triples (z_1, z_2, z_3) and (w_1, w_2, w_3) of distinct points of $\overline{\mathbb{C}}$, suppose there are two elements m and n of Möb^+ satisfying $n(z_1) = w_1 = m(z_1)$, $n(z_2) = w_2 = m(z_2)$, and $n(z_3) = w_3 = m(z_3)$. By Theorem 2.6, we know that since $m^{-1} \circ n$ fixes three distinct points of $\overline{\mathbb{C}}$, it is the identity, and so $m = n$. This completes the proof of uniqueness.

In order to demonstrate the existence of a Möbius transformation taking (z_1, z_2, z_3) to (w_1, w_2, w_3), it suffices to show that there is a Möbius transformation m satisfying $m(z_1) = 0$, $m(z_2) = 1$, and $m(z_3) = \infty$. If we can construct such an m, we can also construct a Möbius transformation n satisfying $n(w_1) = 0$, $n(w_2) = 1$, and $n(w_3) = \infty$, and then $n^{-1} \circ m$ is the desired transformation taking (z_1, z_2, z_3) to (w_1, w_2, w_3).

So, it remains only to construct a Möbius transformation m satisfing $m(z_1) = 0$, $m(z_2) = 1$, and $m(z_3) = \infty$. We work in the case that all the z_k lie in $\mathbb{C}$, and leave the derivation in the case that one of the z_k is ∞ as an exercise. Explicitly, consider the function on $\overline{\mathbb{C}}$ given by

$$m(z) = \frac{z - z_1}{z - z_3} \frac{z_2 - z_3}{z_2 - z_1} = \frac{(z_2 - z_3)z - z_1(z_2 - z_3)}{(z_2 - z_1)z - z_3(z_2 - z_1)}.$$

Just by its construction, we have that $m(z_1) = 0$, $m(z_2) = 1$, and $m(z_3) = \infty$. Moreover, since the z_k are distinct,

$$(z_2 - z_3)(-z_3)(z_2 - z_1) - (-z_1)(z_2 - z_3)(z_2 - z_1) = (z_2 - z_3)(z_1 - z_3)(z_2 - z_1) \neq 0,$$

and so m is actually a Möbius transformation.

Exercise 2.8

Derive the general form of the Möbius transformation taking the triple (∞, z_2, z_3) to the triple $(0, 1, \infty)$.

As is often the case, the actual construction of the specific Möbius transformation taking one triple to another can be fairly unpleasant. For example, let us consider the two triples $(2i, 1+i, 3)$ and $(0, 2+2i, 4)$ and construct the Möbius transformation taking $(2i, 1+i, 3)$ to $(0, 2+2i, 4)$. A warning: this example has not been chosen for its numerical elegance.

Following the proof of existence, we construct the Möbius transformation m taking $(2i, 1+i, 3)$ to $(0, 1, \infty)$ and the Möbius transformation n taking $(0, 2+2i, 4)$ to $(0, 1, \infty)$.

The Möbius transformation m taking $(2i, 1+i, 3)$ to $(0, 1, \infty)$ is given by

$$m(z) = \frac{(z - 2i)}{(z - 3)} \frac{(1 + i - 3)}{(1 + i - 2i)} = \frac{(-2 + i)z + 2 + 4i}{(1 - i)z - 3 + 3i}.$$

The Möbius transformation n taking $(0, 2+2i, 4)$ to $(0, 1, \infty)$ is given as

$$n(z) = \frac{z}{(z-4)} \frac{(2+2i-4)}{(2+2i)} = \frac{(-2+2i)z}{(2+2i)z - 8 - 8i}.$$

So, the transformation we are looking for is

$$n^{-1} \circ m(z) = \frac{(24+8i)z + 16 - 48i}{(6+6i)z + 4 - 24i}.$$

Up to this point, we've been considering *ordered* triples of distinct points in $\overline{\mathbb{C}}$. If we consider *unordered* triples, and in particular if we ask about the Möbius transformations taking one unordered triple of distinct points to another, the proof of existence goes through without change, but the proof of uniqueness no longer holds.

Exercise 2.9

Consider the unordered triple $T = \{0, 1, \infty\}$ of points of $\overline{\mathbb{C}}$. Determine all the Möbius transformations m satisfying $m(T) = T$.

The action of Möb^+ on the set of triples of distinct points of $\overline{\mathbb{C}}$ is an example of a *group action*.

Definition 2.7

Say that a group G *acts on a set* X if there is a homomorphism from G into the group $\text{bij}(X)$ of bijections of X.

That is, a group G acts on a set X if every element of g gives rise to a bijection of X, and moreover if multiplication of elements of G using the group operation corresponds to composition of the corresponding homeomorphisms.

We do not do much with group actions in this book, other than making use of some of the basic terminology. For more information, the interested reader should pick up a book on Abstract Algebra, such as Herstein [12]. Philosophically, considering group actions allows one to view a group not as an abstract object, but as a well-behaved collection of symmetries of a set X.

There are many adjectives that one can apply to group actions, and restrictions that one can apply to the types of bijections considered.

For instance, say that G acts *transitively* on X if for each pair x and y of elements of X, there exists some element g of G satisfying $g(x) = y$. This is one of the properties of most interest to us. The following lemma gives a slightly easier condition to check to obtain transitivity, which is merely a generalization of the idea used in the proof that Möb^+ acts transitively on triples of distinct points of $\overline{\mathbb{C}}$.

Lemma 2.8

Suppose that a group G acts on a set X, and let x_0 be a point of X. Suppose that for each point y of X there exists an element g of G so that $g(y) = x_0$. Then, G acts transitively on X.

Given two points y and z of X, choose elements g_y and g_z of G so that $g_y(y) = x_0 = g_z(z)$. Then, $(g_z)^{-1} \circ g_y(y) = z$. This completes the proof of Lemma 2.8.

Also inspired by considering the action of Möb on triples of distinct points of $\overline{\mathbb{C}}$, say that a group G acts *uniquely transitively* on a set X if for each pair x and y of elements of X, there exists one and only one element g of G with $g(x) = y$. In this language, we can restate what we know about the action of Möb on triples of distinct points of $\overline{\mathbb{C}}$.

Theorem 2.9

Möb^+ acts uniquely transitively on the set $\mathcal{T}$ of triples of distinct points of $\overline{\mathbb{C}}$.

There are other sets of objects in $\overline{\mathbb{C}}$ on which Möb^+ acts transitively.

Theorem 2.10

Möb^+ acts transitively on the set $\mathcal{C}$ of circles in $\overline{\mathbb{C}}$.

The first step in proving Theorem 2.10 is to observe that a triple of distinct points in $\overline{\mathbb{C}}$ determines a unique circle in $\overline{\mathbb{C}}$.

To see this, let (z_1, z_2, z_3) be a triple of distinct points of $\overline{\mathbb{C}}$. If all the z_k lie in $\mathbb{C}$ and are not colinear, then there exists a unique Euclidean circle passing through all three. If all the z_k lie in $\mathbb{C}$ and are colinear, then there exists a unique Euclidean line passing through all three. If one of the z_k is ∞, then there is a unique Euclidean line passing through the other two.

However, while each triple of distinct points of $\overline{\mathbb{C}}$ determines a unique circle in $\overline{\mathbb{C}}$, the converse is not true. Given a circle A in $\overline{\mathbb{C}}$, there are lots and lots of triples of distinct points of $\overline{\mathbb{C}}$ that give rise to A.

So, let A and B be two circles in $\overline{\mathbb{C}}$. Choose a triple of distinct points on A, a triple of distinct points on B, and let m be the Möbius transformation taking the triple of points determining A to the triple of points determining B. Since $m(A)$ and B are then two circles in $\overline{\mathbb{C}}$ that pass through the same triple of points, we have that $m(A) = B$. This completes the proof of Theorem 2.10.

However, the fact that a circle in $\overline{\mathbb{C}}$ does not determine a unique triple of distinct points in $\overline{\mathbb{C}}$ means that this action is not uniquely transitive. That is, given two circles in $\overline{\mathbb{C}}$, there are in fact many Möbius transformations taking one to the other.

For example, we may mimic Exercise 2.9. Let (z_1, z_2, z_3) be a triple of distinct points, and let A be the circle in $\overline{\mathbb{C}}$ determined by (z_1, z_2, z_3). Then, the identity takes A to A. However, the Möbius transformation taking (z_1, z_2, z_3) to (z_2, z_1, z_3) also takes A to A. We will encounter this phenomenon again in Section 2.8, in which we determine the set of Möbius transformations taking any circle A in $\overline{\mathbb{C}}$ to itself.

We can rephrase this argument as saying that there exists a well-defined surjective function from the set $\mathcal{T}$ of triples of distinct points of $\overline{\mathbb{C}}$ to the set $\mathcal{C}$ of circles in $\overline{\mathbb{C}}$. Since Möb^+ acts transitively on $\mathcal{T}$, we can use the function from $\mathcal{T}$ to $\mathcal{C}$ to push down the action of Möb^+ from $\mathcal{T}$ to $\mathcal{C}$.

We can also consider the action of Möb^+ on the set $\mathcal{D}$ of discs in $\overline{\mathbb{C}}$.

Theorem 2.11

Möb^+ acts transitively on the set $\mathcal{D}$ of discs in $\overline{\mathbb{C}}$.

As might be expected, the proof of Theorem 2.11 is very similar to the proof of Theorem 2.10. In fact, the proofs differ in only one respect.

Let D and E be two discs in $\overline{\mathbb{C}}$, where D is determined by the circle C_D in $\overline{\mathbb{C}}$ and E is determined by the circle C_E in $\overline{\mathbb{C}}$. Since Möb^+ acts transitively on the set $\mathcal{C}$ of circles in $\overline{\mathbb{C}}$, there is a Möbius transformation m satisfying $m(C_D) = C_E$, and so $m(D)$ is a disc determined by C_E.

However, there are two discs determined by C_E, and we have no way of knowing whether $m(D) = E$ or whether $m(D)$ is the other disc determined by C_E. If $m(D) = E$, we are done. If $m(D) \neq E$, we need to find a Möbius transformation taking C_E to itself and interchanging the two discs determined by C_E.

This is not too difficult. We first work with a circle in $\overline{\mathbb{C}}$ we understand, and then use the transitivity of Möb^+ on the set of circles in $\overline{\mathbb{C}}$ to transport our solution for this particular circle to any other circle.

For the circle $\overline{\mathbb{R}}$, we have already seen the answer to this question, namely the Möbius transformation $J(z) = \frac{1}{z}$. Since $J(0) = \infty$, $J(\infty) = 0$, and $J(1) = 1$, we see that J takes $\overline{\mathbb{R}}$ to itself. Since $J(i) = \frac{1}{i} = -i$, we see that J does not take $\mathbb{H}$ to itself, and so J interchanges the two discs determined by $\overline{\mathbb{R}}$.

Now, let A be any circle in $\overline{\mathbb{C}}$ and let n be a Möbius transformation satisfying $n(A) = \overline{\mathbb{R}}$. Then, the Möbius transformation $n^{-1} \circ J \circ n$ takes A to itself and interchanges the two discs determined by A. This completes the proof of Theorem 2.11.

As with determining the Möbius transformation taking one triple of distinct points of $\overline{\mathbb{C}}$ to another triple of distinct points, it can be somewhat messy to write out the Möbius transformation taking one disc in $\overline{\mathbb{C}}$ to another.

Consider the two discs

$$D = \{z \in \mathbb{C} : |z| < 2\} \text{ and } E = \{z \in \mathbb{C} : |z - (4 + 5i)| < 1\}.$$

There are many different Möbius transformations taking E to D. We construct one.

Let $m(z) = z - 4 - 5i$. Since E is the Euclidean disc with centre $4 + 5i$ and radius 1, we have that $m(E)$ is the Euclidean disc with centre $m(4 + 5i) = 0$ and radius 1. If we now compose m with $n(z) = 2z$, we see that $n \circ m(E)$ is the Euclidean disc with centre 0 and radius 2, so that $n \circ m(E) = D$, as desired. Writing out $n \circ m$ explicitly, we get $n \circ m(z) = n(z - 4 - 5i) = 2z - 8 - 10i$.

Exercise 2.10

Give an explicit Möbius transformation taking $\mathbb{D}$ to $\mathbb{H}$.

2.3 The Cross Ratio

In Section 2.2, we considered the transitivity properties of Möb^+. We saw that Möb^+ acts uniquely transitively on the set $\mathcal{T}$ of ordered triples of distinct points of $\overline{\mathbb{C}}$, and acts transitively on both the set $\mathcal{C}$ of circles in $\overline{\mathbb{C}}$ and the set $\mathcal{D}$ of discs in $\overline{\mathbb{C}}$.

In this section, we consider a different sort of question, and start to ask about functions on $\overline{\mathbb{C}}$ that are invariant under Möb^+. Variants of this question will occupy our attention at different times throughout the book.

Definition 2.12

By a *function invariant under* Möb^+, we mean a function f of variables $z_1, \ldots, z_k$, where each z_k lies in $\overline{\mathbb{C}}$, so that

$$f(z_1, \ldots, z_k) = f(m(z_1), \ldots, m(z_k))$$

for all $m \in \text{Möb}^+$.

Exercise 2.11

Show that the function $f : \overline{\mathbb{C}} \to \overline{\mathbb{C}}$ given by $f(z) = z^2$ for $z \in \mathbb{C}$ and $f(\infty) = \infty$, is not invariant under Möb^+. Determine whether there exists a subgroup of Möb^+ under which f is invariant.

In fact, the triple transitivity of Möb^+ on $\overline{\mathbb{C}}$ implies that for $1 \le n \le 3$, the only functions $f : \overline{\mathbb{C}}^n \to \overline{\mathbb{C}}$ invariant under Möb^+ are the constant functions.

For functions $f : \overline{\mathbb{C}}^n \to \overline{\mathbb{C}}$ for $n \ge 4$, the situation becomes more interesting. One example of a function of 4 variables on $\overline{\mathbb{C}}$ invariant under Möb^+ is the cross ratio.

Definition 2.13

Given four distinct points z_1, z_2, z_3, and z_4 in $\mathbb{C}$, define the *cross ratio* of z_1, z_2, z_3, and z_4 to be

$$[z_1, z_2; z_3, z_4] = \frac{(z_1 - z_4)}{(z_1 - z_2)} \frac{(z_3 - z_2)}{(z_3 - z_4)}.$$

Following our usual pattern, if one of the z_k is ∞, we define the cross ratio by continuity. That is, we set

$$\begin{aligned}[\infty, z_2; z_3, z_4] = \lim_{z \to \infty} (z, z_2; z_3, z_4) &= \lim_{z \to \infty} \frac{(z - z_4)}{(z - z_2)} \frac{(z_3 - z_2)}{(z_3 - z_4)} \\ &= \lim_{z \to \infty} \frac{(1 - \frac{z_4}{z})}{(1 - \frac{z_2}{z})} \frac{(z_3 - z_2)}{(z_3 - z_4)} = \frac{z_3 - z_2}{z_3 - z_4}.\end{aligned}$$

The cross ratios $[z_1, \infty; z_3, z_4]$, $[z_1, z_2; \infty, z_4]$, and $[z_1, z_2; z_3, \infty]$ are defined similarly.

Exercise 2.12

Show that the cross ratio is invariant under Möb^+.

There are some cases in which the cross ratio is particularly easy to calculate. For example, consider $[\infty, 0; 1, z]$. From what we have done above, we have that

$$[\infty, 0; 1, z] = \frac{1}{1-z} = \frac{1-\overline{z}}{|1-z|^2}.$$

In particular, we have that $[\infty, 0; 1, z]$ is real if and only if $\overline{z}$, and hence z, is real.

Combining this with the fact that the cross ratio is invariant under Möb^+ gives us a very easy test to see whether four distinct points of $\overline{\mathbb{C}}$ lie on a circle in $\overline{\mathbb{C}}$.

Proposition 2.14

Let z_1, z_2, z_3, and z_4 be four distinct points in $\overline{\mathbb{C}}$. Then, z_1, z_2, z_3, and z_4 lie on a circle in $\overline{\mathbb{C}}$ if and only if the cross ratio $[z_1, z_2; z_3, z_4]$ is real.

Let z_1, z_2, z_3, and z_4 be four distinct points in $\overline{\mathbb{C}}$, and let m be a Möbius transformation satisfying $m(z_1) = \infty$, $m(z_2) = 0$, and $m(z_3) = 1$.

Observe that $m(z_1) = \infty$, $m(z_2) = 0$, $m(z_3) = 1$, and $m(z_4)$ lie on a circle in $\overline{\mathbb{C}}$, namely $\overline{\mathbb{R}}$, if and only if $m(z_4)$, and hence $[m(z_1), m(z_2); m(z_3), m(z_4)]$, is real.

Since $[z_1, z_2; z_3, z_4] = [m(z_1), m(z_2); m(z_3), m(z_4)]$ and since Möb^+ takes circles in $\overline{\mathbb{C}}$ to circles in $\overline{\mathbb{C}}$, we have that z_1, z_2, z_3, and z_4 lie on a circle in $\overline{\mathbb{C}}$ if and only if $[z_1, z_2; z_3, z_4]$ is real. This completes the proof of Proposition 2.14.

Exercise 2.13

Determine whether $2+3i$, $-2i$, $1-i$, and 4 lie on a circle in $\overline{\mathbb{C}}$.

Exercise 2.14

Determine the real values of s for which the points $2+3i$, $-2i$, $1-i$, and s lie on a circle in $\overline{\mathbb{C}}$.

There is some amount of choice inherent in the definition of the cross ratio. For instance, we can also consider the cross ratios

$$[z_1, z_2; z_3, z_4]_2 = \frac{(z_1 - z_2)}{(z_1 - z_4)} \frac{(z_3 - z_4)}{(z_3 - z_2)}$$

and

$$[z_1, z_2; z_3, z_4]_3 = \frac{(z_2 - z_1)}{(z_2 - z_4)} \frac{(z_3 - z_4)}{(z_3 - z_1)}.$$

We note here that all the possible choices of cross ratio, such as the three described in this section, are all closely related.

Exercise 2.15

Express the two cross ratios $[z_1, z_2; z_3, z_4]_2$ and $[z_1, z_2; z_3, z_4]_3$ in terms of the standard cross ratio $[z_1, z_2; z_3, z_4]$.

2.4 Classification of Möbius Transformations

The classification of Möbius transformations given in Section 2.1, in terms of the number of fixed points, is as we wrote at the time very crude and can be considerably refined.

Before getting into the refinement of this classification, we introduce a notion of sameness for Möbius transformations.

Definition 2.15

Say that two Möbius transformations m_1 and m_2 are *conjugate* if there exists some Möbius transformation p so that $m_2 = p \circ m_1 \circ p^{-1}$.

Geometrically, if m_1 and m_2 are conjugate by p, then the action of m_1 on $\overline{\mathbb{C}}$ is the same as the action of m_2 on $p(\overline{\mathbb{C}}) = \overline{\mathbb{C}}$. That is, conjugacy reflects a change of coordinates on $\overline{\mathbb{C}}$.

Exercise 2.16

Suppose that m and n are Möbius transformations that are conjugate by p, so that $m = p \circ n \circ p^{-1}$. Prove that m and n have the same number of fixed points in $\overline{\mathbb{C}}$.

The basic idea of the classification of Möbius transformations is to conjugate a given Möbius transformation into a standard form, and then classify the possible standard forms. For the remainder of this section, we work with a Möbius transformation m that is not the identity.

Suppose that m has only one fixed point in $\overline{\mathbb{C}}$, and call it x. Let y be any point of $\overline{\mathbb{C}} - \{x\}$, and observe that $(x, y, m(y))$ is a triple of distinct points of $\overline{\mathbb{C}}$.

Let p be the Möbius transformation taking the triple $(x, y, m(y))$ to the triple $(\infty, 0, 1)$, and consider the composition $p \circ m \circ p^{-1}$.

By our construction of p, we have that $p \circ m \circ p^{-1}(\infty) = p \circ m(x) = p(x) = \infty$. Since $p \circ m \circ p^{-1}$ fixes ∞, we can write it as $p \circ m \circ p^{-1}(z) = az + b$ with $a \neq 0$. Since $p \circ m \circ p^{-1}$ has only the one fixed point in $\overline{\mathbb{C}}$, namely ∞, there is no solution in $\mathbb{C}$ to the equation $p \circ m \circ p^{-1}(z) = z$, and so it must be that $a = 1$.

Since $p \circ m \circ p^{-1}(0) = p \circ m(y) = 1$, we see that $b = 1$ as well, and so $p \circ m \circ p^{-1}(z) = z + 1$. Therefore, any Möbius transformation m with only one fixed point is conjugate by a Möbius transformation to $n(z) = z + 1$. We say that m is *parabolic*, and we refer to $p \circ m \circ p^{-1}(z) = z + 1$ as its *standard form.*

To consider a specific example, let $m(z) = \frac{z}{z+1}$. Since $m(\infty) = 1 \neq \infty$, the fixed points of m are the solutions in $\mathbb{C}$ to the equation $m(z) = \frac{z}{z+1} = z$, which are the solutions to $z = z^2 + z$. Hence, the only fixed point of m is 0.

To find the Möbius transformation p conjugating m to its standard form, choose some point in $\overline{\mathbb{C}} - \{0\}$, say ∞, and calculate that $m(\infty) = 1$. Then, we take p to be the Möbius transformation sending the triple $(0, \infty, 1)$ to the triple $(\infty, 0, 1)$, namely $p(z) = \frac{iz}{i} = \frac{1}{z}$.

In the argument just given, there is some ambiguity in the choice of the conjugating Möbius transformation p, as the specific form of p depends on the choice of the point y not fixed by m. However, this choice does not play an essential role.

Suppose now that m has two fixed points in $\overline{\mathbb{C}}$, and call them x and y. Let q be a Möbius transformation satisfying $q(x) = 0$ and $q(y) = \infty$, and consider the composition $q \circ m \circ q^{-1}$.

By definition, we have that $q \circ m \circ q^{-1}(\infty) = q \circ m(y) = q(y) = \infty$, and that $q \circ m \circ q^{-1}(0) = q \circ m(x) = q(x) = 0$, and so we may write $q \circ m \circ q^{-1}(z) = az$ for some $a \in \mathbb{C} - \{0,\ 1\}$. We refer to a as the *multiplier of m.*

To consider a specific example, let $m(z) = \frac{2z+1}{z+1}$. Since $m(\infty) = 2 \neq \infty$, the fixed points of m are the solutions in $\mathbb{C}$ to the equation $m(z) = \frac{2z+1}{z+1} = z$, which are the solutions to $z^2 - z - 1 = 0$. Using the quadratic formula, we see that the fixed points of m are $z = \frac{1}{2}(1 \pm \sqrt{5})$.

To find the Möbius transformation q conjugating m to its standard form, consider a Möbius transformation q taking $\frac{1}{2}(1 + \sqrt{5})$ to 0 and taking $\frac{1}{2}(1 - \sqrt{5})$ to ∞, for instance

$$q(z) = \frac{z - \frac{1}{2}(1 + \sqrt{5})}{z - \frac{1}{2}(1 - \sqrt{5})}.$$

At this point, instead of calculating out the composition $q \circ m \circ q^{-1}$ explicitly, which we can certainly do, we calculate the multiplier of m by calculating the single value

$$a = q \circ m \circ q^{-1}(1) = q \circ m(\infty) = q(2) = \frac{3-\sqrt{5}}{3+\sqrt{5}}.$$

As in the argument for parabolic Möbius transformations, there is some ambiguity in the choice of the conjugating Möbius transformation q, as there is not enough information to specify q uniquely. However, as in seen in the following two exercises, this choice does not play an essential role.

Exercise 2.17

Let m be a Möbius transformation with two fixed points x and y. Prove that if n_1 and n_2 are two Möbius transformations satisfying $n_1(x) = 0 = n_2(x)$ and $n_1(y) = \infty = n_2(y)$, then the multipliers of $n_1 \circ m \circ n_1^{-1}$ and $n_2 \circ m \circ n_2^{-1}$ are equal.

Exercise 2.18

Using the notation of the argument just given for Möbius transformations with two fixed points, prove that if we conjugate m as above by a Möbius transformation s satisfying $s(x) = \infty$ and $s(y) = 0$, the multiplier of $s \circ m \circ s^{-1}$ is $\frac{1}{a}$.

As a corollary to Exercises 2.17 and 2.18, we see that the multiplier of a Möbius transformation with two fixed points is only defined up to taking its inverse. Moreover, the solution to Exercise 2.18 shows that $J(z) = \frac{1}{z}$ conjugates $m(z) = az$ to $m^{-1}(z) = \frac{1}{a}z$.

If the multiplier of m satisfies $|a| = 1$, then we may write $a = e^{2i\varphi}$ for some φ in $(0, \pi)$, and $q \circ m \circ q^{-1}(z) = e^{2i\varphi}z$ is rotation about the origin by angle 2φ. We say that m is *elliptic*, and we refer to $q \circ m \circ q^{-1}(z) = e^{2i\varphi}z$ as its *standard form.*

If on the other hand $|a| \neq 1$, then we may write $a = \lambda^2 e^{2i\varphi}$ for some positive real number $\lambda \neq 1$ and some φ in $[0, \pi)$, so that $q \circ m \circ q^{-1}(z) = \lambda^2 e^{2i\varphi}z$ is the composition of a dilation by λ^2 (an expansion if $\lambda^2 > 1$ or a contraction if $\lambda^2 < 1$) and a (possibly trivial) rotation about the origin by angle 2φ. We say that m is *loxodromic*, and we refer to $q \circ m \circ q^{-1}(z) = \lambda^2 e^{2i\varphi}z$ as its *standard form.*

Exercise 2.19

Determine the type, parabolic, elliptic, or loxodromic, of each of the Möbius transformations given in Exercise 2.7.

The name loxodromic comes from the word *loxodrome*, which is a curve on the sphere that meets every line of latitude at the same angle. Lines of longitude are loxodromes, but there are also loxodromes that spiral into both poles. The reason these Möbius transformations are called loxodromic is that each one keeps invariant a loxodrome.

2.5 A Matrix Representation

If we examine the formula for the composition of two Möbius transformations, we get a hint that there is a strong connection between Möbius transformations and 2×2 matrices. Consider the Möbius transformations $m(z) = \frac{az+b}{cz+d}$ and $n(z) = \frac{\alpha z+\beta}{\gamma z+\delta}$. Then,

$$\begin{aligned} n \circ m(z) = \frac{\alpha m(z) + \beta}{\gamma m(z) + \delta} &= \frac{\alpha\left(\frac{az+b}{cz+d}\right) + \beta}{\gamma\left(\frac{az+b}{cz+d}\right) + \delta} \\ &= \frac{\alpha(az+b) + \beta(cz+d)}{\gamma(az+b) + \delta(cz+d)} \\ &= \frac{(\alpha a + \beta c)z + \alpha b + \beta d}{(\gamma a + \delta c)z + \gamma b + \delta d}. \end{aligned}$$

If instead we view the coefficients of m and n as the entries in a pair of 2×2 matrices, we get

$$\begin{pmatrix} \alpha & \beta \\ \gamma & \delta \end{pmatrix} \begin{pmatrix} a & b \\ c & d \end{pmatrix} = \begin{pmatrix} \alpha a + \beta c & \alpha b + \beta d \\ \gamma a + \delta c & \gamma b + \delta d \end{pmatrix},$$

and the entries of the product matrix correspond to the entries of the composition of the two Möbius transformations.

We will examine the details of this correspondence between Möbius transformations and matrices later in the section. For the moment, let us concentrate on using this similarity to refine still further the classification of Möbius transformations we discussed in Section 2.4.

There are two main numerical quantities one can associate to a 2×2 matrix, the *determinant* and the *trace*. Using this correspondence between matrices and

Möbius transformations, we can define similar notions for Möbius transformations.

Define the *determinant* $\det(m)$ of the Möbius transformation $m(z) = \frac{az+b}{cz+d}$ to be the quantity $\det(m) = ad - bc$. Note that the determinant of a Möbius transformation is not a well-defined quantity. If we multiply the coefficients of m by any non-zero constant, this has no effect on the action of m on $\overline{\mathbb{C}}$, since

$$\frac{az+b}{cz+d} = \frac{\alpha az + \alpha b}{\alpha cz + \alpha d}$$

for all $\alpha \in \mathbb{C} - \{0\}$ and all $z \in \overline{\mathbb{C}}$. However, the determinants are not equal, since the determinant of $f(z) = \frac{az+b}{cz+d}$ is $\det(f) = ad - bc$ and the determinant of $g(z) = \frac{\alpha az+\alpha b}{\alpha cz+\alpha d}$ is $\det(g) = \alpha^2(ad - bc)$.

Exercise 2.20

Calculate the determinants of the following Möbius transformations:

1. $m(z) = \frac{2z+4}{5z-7}$; 2. $m(z) = \frac{1}{z}$; 3. $m(z) = \frac{-z-3}{z+1}$;

4. $m(z) = \frac{iz+1}{z+3i}$; 5. $m(z) = iz + 1$; 6. $m(z) = \frac{-z}{z+4}$;

However, we can always choose α so that the determinant of m is 1. This still leaves a small amount of ambiguity, since all the coefficients of m can all be multiplied by -1 without changing the determinant of m, but this is the only remaining ambiguity. We refer to this process as *normalizing* m.

Exercise 2.21

Normalize each of the Möbius transformations from Exercise 2.20.

Having normalized a Möbius transformation m, there is another useful numerical quantity associated to m, which corresponds to taking the trace. Consider the function

$$\tau : \text{Möb}^+ \to \mathbb{C}$$

defined by setting $\tau(m) = (a+d)^2$, where $m(z) = \frac{az+b}{cz+d}$. Since the only ambiguity in the definition of a normalized Möbius transformation arises from multiplying all the coefficients by -1, we see that $\tau(m)$ is well-defined. In fact, this possible ambiguity is why we consider the function τ and not the actual trace $\text{trace}(m) = a + d$.

As with the trace of a matrix, one very useful property of τ is that it is invariant under conjugation.

Exercise 2.22

Show that $\tau(m \circ n) = \tau(n \circ m)$.

Exercise 2.23

Show that $\tau(p \circ m \circ p^{-1}) = \tau(m)$.

Using this invariance of τ under conjugation, we are able to distinguish the different types of Möbius transformations without explicitly conjugating them to their standard forms. Namely, let m be a Möbius transformation, and let p be a Möbius transformation conjugating m to its standard form. Since $\tau(m) = \tau(p \circ m \circ p^{-1})$, it suffices to consider the values of τ on the standard forms.

If m is parabolic, then $p \circ m \circ p^{-1}(z) = z + 1$, and so

$$\tau(m) = \tau(p \circ m \circ p^{-1}) = (1 + 1)^2 = 4.$$

Note that, for the identity Möbius transformation $e(z) = z$ we also have that $\tau(e) = (1 + 1)^2 = 4$.

If m is either elliptic or loxodromic, we may write $p \circ m \circ p^{-1}(z) = \alpha^2 z$, where $\alpha^2 \in \mathbb{C} - \{0, 1\}$. Normalizing so that the determinant of m is 1 yields that we need to write

$$m(z) = \frac{\alpha z}{\alpha^{-1}},$$

and so

$$\tau(m) = \tau(p \circ m \circ p^{-1}) = (\alpha + \alpha^{-1})^2.$$

In the case that m is elliptic, so that $|\alpha| = 1$, write $\alpha = e^{i\theta}$ for some θ in $(0, \pi)$. Calculating, we see that

$$\tau(m) = (\alpha + \alpha^{-1})^2 = \left(e^{i\theta} + e^{-i\theta}\right)^2 = 4\cos^2(\theta).$$

In particular, we have that $\tau(m)$ is real and lies in the interval $[0, 4)$.

In the case that m is loxodromic, so that $|\alpha| \neq 1$, we write $\alpha = \rho e^{i\theta}$ for some $\rho > 0$, $\rho \neq 1$, and some θ in $[0, \pi)$. Calculating, we see that

$$\alpha + \alpha^{-1} = \rho e^{i\theta} + \rho^{-1} e^{-i\theta},$$

and so

$$\tau(m) = (\alpha + \alpha^{-1})^2 = \cos(2\theta)(\rho^2 + \rho^{-2}) + 2 + i\sin(2\theta)(\rho^2 - \rho^{-2}).$$

In particular, since $\rho \neq 1$, we see that $\mathrm{Im}(\tau(m)) \neq 0$ for $\theta \neq 0$ and $\theta \neq \frac{\pi}{2}$.

For the two cases that $\theta = 0$ and $\theta = \frac{\pi}{2}$, we use the following exercise from Calculus.

Exercise 2.24

Show that the function $f : (0,\infty) \to \mathbb{R}$ defined by $f(\rho) = \rho^2 + \rho^{-2}$ satisfies $f(\rho) \geq 2$, with $f(\rho) = 2$ if and only if $\rho = 1$.

For $\theta = 0$, we see that $\tau(m) > 4$, while for $\theta = \frac{\pi}{2}$, we see that $\tau(m) < 0$.

To summarize, we have shown the following.

Proposition 2.16

Let m be a Möbius transformation other than the identity. Then,

1 m is parabolic if and only if $\tau(m) = 4$;

2 m is elliptic if and only if $\tau(m)$ is real and lies in $[0, 4)$;

3 m is loxodromic if and only if either $\tau(m)$ has non-zero imaginary part, or $\tau(m)$ is real and lies in $(-\infty, 0) \cup (4, \infty)$.

To work through a specific example, consider $m(z) = \frac{z+1}{z+3}$. The determinant of m is $3 - 1 = 2$, and so the normalized form of m is

$$m(z) = \frac{\frac{1}{\sqrt{2}}z + \frac{1}{\sqrt{2}}}{\frac{1}{\sqrt{2}}z + \frac{3}{\sqrt{2}}}.$$

Calculating, we see that $\tau(m) = 8$, and so m is loxodromic.

Note that we are able to determine the multiplier of an elliptic or loxodromic transformation m, up to taking its inverse, knowing only the value of $\tau(m)$. Specifically, if m has multiplier λ^2, then

$$\tau(m) = (\lambda + \lambda^{-1})^2 = \lambda^2 + \lambda^{-2} + 2.$$

Multiplying through by λ^2 gives

$$\lambda^4 + (2 - \tau(m))\lambda^2 + 1 = 0.$$

Applying the quadratic formula, we obtain

$$\begin{aligned}\lambda^2 &= \frac{1}{2}\left[\tau(m) - 2 \pm \sqrt{(2-\tau(m))^2 - 4}\right] \\ &= \frac{1}{2}\left[\tau(m) - 2 \pm \sqrt{-4\tau(m) + \tau^2(m)}\right].\end{aligned}$$

Since

$\frac{1}{2}\left[\tau(m) - 2 + \sqrt{-4\tau(m) + \tau^2(m)}\right]$ and $\frac{1}{2}\left[\tau(m) - 2 - \sqrt{-4\tau(m) + \tau^2(m)}\right]$

are inverses of one another, we may take the multiplier λ^2 to satisfy $|\lambda|^2 > 1$.

Exercise 2.25

Determine the type of each of the Möbius transformations from Exercise 2.20. If the transformation is elliptic or loxodromic, determine its multiplier.

Exercise 2.26

Show that if m is a parabolic Möbius transformation with fixed point $x \neq \infty$, then there exists a unique complex number p so that

$$m(z) = \frac{(1+px)z - px^2}{pz + 1 - px}.$$

Exercise 2.27

Show that if m is a Möbius transformation with distinct fixed points $x \neq \infty$ and $y \neq \infty$ and multiplier a, then we can write

$$m(z) = \frac{\left(\frac{x-ya}{x-y}\right) z + \frac{xy(a-1)}{x-y}}{\left(\frac{1-a}{x-y}\right) z + \frac{xa-y}{x-y}}.$$

We close this section by making explicit the correspondence between Möbius transformations and 2×2 matrices. To set notation, let

$$\mathrm{GL}_2(\mathbb{C}) = \left\{ \left(\begin{array}{cc} a & b \\ c & d \end{array} \right) \mid a, b, c, d \in \mathbb{C} \text{ and } ad - bc \neq 0 \right\},$$

and let

$$\mathrm{SL}_2(\mathbb{C}) = \left\{ \left(\begin{array}{cc} a & b \\ c & d \end{array} \right) \mid a, b, c, d \in \mathbb{C} \text{ and } ad - bc = 1 \right\},$$

We have already seen, in our discussion of normalization, that a Möbius transformation determines many matrices, so the obvious guess of a function from $\mathrm{Möb}^+$ to $\mathrm{GL}_2(\mathbb{C})$ is not well-defined.

So we go the other way and consider the obvious choice of a function from $\mathrm{GL}_2(\mathbb{C})$ to $\mathrm{Möb}^+$. Define $\mu : \mathrm{GL}_2(\mathbb{C}) \to \mathrm{Möb}^+$ by

$$\mu\left(M = \left(\begin{array}{cc} a & b \\ c & d \end{array} \right) \right) = \left(m(z) = \frac{az+b}{cz+d} \right).$$

Note that the calculation done at the beginning of this section proves that μ is a homomorphism.

Exercise 2.28

Prove that the kernel $\ker(\mu)$ of μ is the subgroup $K = \{\lambda \mathrm{I} \mid \lambda \in \mathbb{C}\}$ of $\mathrm{GL}_2(\mathbb{C})$. Conclude that Möb^+ is isomorphic to $\mathrm{PGL}_2(\mathbb{C}) = \mathrm{GL}_2(\mathbb{C})/K$.

2.6 Reflections

We have seen, in Theorem 2.5, that Möb^+ is contained in the set $\text{Homeo}^{\mathrm{C}}(\overline{\mathbb{C}})$ of homeomorphisms of $\overline{\mathbb{C}}$ that take circles in $\overline{\mathbb{C}}$ to circles in $\overline{\mathbb{C}}$. There is a natural extension of Möb^+ that also lies in $\text{Homeo}^{\mathrm{C}}(\overline{\mathbb{C}})$.

In order to extend Möb^+ to a larger group, we consider the simplest homeomorphism of $\overline{\mathbb{C}}$ not already in Möb^+, namely *complex conjugation*. Set

$$C(z) = \overline{z} \text{ for } z \in \mathbb{C} \text{ and } C(\infty) = \infty.$$

Proposition 2.17

The function $C : \overline{\mathbb{C}} \to \overline{\mathbb{C}}$ defined by

$$C(z) = \overline{z} \text{ for } z \in \mathbb{C} \text{ and } C(\infty) = \infty$$

is an element of $\text{Homeo}(\overline{\mathbb{C}})$.

Note that C is its own inverse, that is $C^{-1}(z) = C(z)$, and hence C is a bijection of $\overline{\mathbb{C}}$. So, we need only check that C is continuous.

The continuity of C follows from the observation that for any point z of $\overline{\mathbb{C}}$ and any $\varepsilon > 0$, we have that $C(U_\varepsilon(z)) = U_\varepsilon(C(z))$. This completes the proof of Proposition 2.17.

Exercise 2.29

Show that C is not an element of Möb^+.

Definition 2.18

The *general Möbius group* Möb is the group generated by Möb^+ and C. That is, every (non-trivial) element p of Möb can be expressed as a composition

$$p = C \circ m_k \circ \cdots C \circ m_1$$

for some $k \geq 1$, where each m_k is an element of Möb^+.

Note that since Möb contains Möb^+, all the transitivity properties of Möb^+ discussed in Section 2.2 are inherited by Möb. That is, Möb acts transitively on the set $\mathcal{T}$ of triples of distinct points in $\overline{\mathbb{C}}$, on the set $\mathcal{C}$ of circles in $\overline{\mathbb{C}}$, and on the set $\mathcal{D}$ of discs in $\overline{\mathbb{C}}$.

Unfortunately, though, Möb does not inherit unique transitively on triples of distinct points, as we saw in the solution to Exercise 2.29.

The proof that $C : \overline{\mathbb{C}} \to \overline{\mathbb{C}}$ lies in $\text{Homeo}^{\text{C}}(\overline{\mathbb{C}})$ is very much like the proof that the elements of Möb^+ lie in $\text{Homeo}^{\text{C}}(\overline{\mathbb{C}})$.

Exercise 2.30

Show that the function $C : \overline{\mathbb{C}} \to \overline{\mathbb{C}}$ lies in $\text{Homeo}^{\text{C}}(\overline{\mathbb{C}})$.

Exercise 2.30 and Theorem 2.5 combine to give the following theorem.

Theorem 2.19

$\text{Möb} \subset \text{Homeo}^{\text{C}}(\overline{\mathbb{C}})$.

We are also able to write down explicit expressions for every element of Möb.

Exercise 2.31

Show that every element of Möb has either the form

$$m(z) = \frac{az+b}{cz+d}$$

or the form

$$n(z) = \frac{a\overline{z}+b}{c\overline{z}+d},$$

where a, b, c, $d \in \mathbb{C}$ and $ad - bc \neq 0$.

Geometrically, the action of C on $\overline{\mathbb{C}}$ is *reflection* in the extended real axis $\overline{\mathbb{R}}$. That is, every point of $\overline{\mathbb{R}}$ is fixed by C, and every point z of $\mathbb{C} - \mathbb{R}$ has the property that $\mathbb{R}$ is the perpendicular bisector of the Euclidean line segment joining z and $C(z)$.

Given that we have defined reflection in the specific circle $\overline{\mathbb{R}}$, and given that Möb acts transitively the set $\mathcal{C}$ of circles in $\overline{\mathbb{C}}$, we may define *reflection* in any circle in $\overline{\mathbb{C}}$.

Specifically, for a circle A in $\overline{\mathbb{C}}$, we choose an element m of Möb taking $\overline{\mathbb{R}}$ to A, and define *reflection in* A to be the composition $C_A = m \circ C \circ m^{-1}$. Note that there is some potential for ambiguity in this definition of C_A, in that there are many choices for the transformation m. We will show in Section 2.8 that C_A is well-defined.

For example, consider $A = \mathbb{S}^1$. One element of Möb^+ taking $\overline{\mathbb{R}}$ to $\mathbb{S}^1$ is the transformation taking the triple $(0, 1, \infty)$ to the triple $(i, 1, -i)$, namely

$$m(z) = \frac{\frac{1}{\sqrt{2}}z + \frac{i}{\sqrt{2}}}{\frac{i}{\sqrt{2}}z + \frac{1}{\sqrt{2}}}.$$

Calculating, we see that

$$C_A(z) = m \circ C \circ m^{-1}(z) = \frac{1}{\overline{z}}.$$

Exercise 2.32

Write down explicit expressions for two distinct elements p and n of Möb taking $\overline{\mathbb{R}}$ to $\mathbb{S}^1$. Show that $p \circ C \circ p^{-1} = n \circ C \circ n^{-1}$.

In the case that A is the Euclidean circle in $\mathbb{C}$ with centre α and radius ρ, we may conjugate reflection in $\mathbb{S}^1$, namely $c(z) = \frac{1}{\overline{z}}$, by the Möbius transformation p taking $\mathbb{S}^1$ to A, namely $p(z) = \rho z + \alpha$, to obtain an explicit expression for the reflection C_A in A, namely

$$C_A(z) = p \circ c \circ p^{-1}(z) = \frac{\rho^2}{\overline{z} - \overline{\alpha}} + \alpha.$$

Similarly, if A is the Euclidean line in $\mathbb{C}$ passing through α and making angle θ with $\mathbb{R}$, we may conjugate reflection in $\mathbb{R}$, namely $C(z) = \overline{z}$, by the Möbius transformation p taking $\mathbb{R}$ to A, namely $p(z) = e^{i\theta}z + \alpha$, to obtain an explicit expression for the reflection C_A in A, namely

$$C_A(z) = p \circ C \circ p^{-1}(z) = e^{2i\theta}(\overline{z} - \overline{\alpha}) + \alpha.$$

This construction of reflections in circles in $\overline{\mathbb{C}}$ has the following consequence.

Proposition 2.20

Every element of Möb can be expressed as the composition of reflections in finitely many circles in $\overline{\mathbb{C}}$.

Since Möb is generated by Möb^+ and $C(z) = \overline{z}$, and since Möb^+ is generated by $J(z) = \frac{1}{z}$ and the $f(z) = az + b$ for $a, b \in \mathbb{C}$ with $a \neq 0$, we need only verify the proposition for these transformations.

By definition, C is reflection in $\overline{\mathbb{R}}$. We can express J as the composition of $C(z) = \overline{z}$ and the reflection $c(z) = \frac{1}{\overline{z}}$ in $\mathbb{S}^1$. Hence, the proof of Proposition 2.20 is completed by the following exercise.

Exercise 2.33

Express every element of Möb^+ of the form $f(z) = az + b$ for $a, b \in \mathbb{C}$ with $a \neq 0$, as the composition of reflections in finitely many circles in $\overline{\mathbb{C}}$.

Over the past several sections, we have seen that the elements of Möb are homeomorphisms of $\overline{\mathbb{C}}$ that take circles in $\overline{\mathbb{C}}$ to circles in $\overline{\mathbb{C}}$. In fact, this property characterizes Möb.

Theorem 2.21

$\text{Möb} = \text{Homeo}^{\text{C}}(\overline{\mathbb{C}})$.

We close this section with a sketch of the proof of Theorem 2.21. By Theorem 2.19, we already have that $\text{Möb} \subset \text{Homeo}^{\text{C}}(\overline{\mathbb{C}})$, and so it remains only to show the opposite inclusion, that $\text{Homeo}^{\text{C}}(\overline{\mathbb{C}}) \subset \text{Möb}$.

So, let f be an element of $\text{Homeo}^{\text{C}}(\overline{\mathbb{C}})$. Let p be the Möbius transformation taking the triple $(f(0), f(1), f(\infty))$ to the triple $(0, 1, \infty)$, so that $p \circ f$ satisfies $p \circ f(0) = 0$, $p \circ f(1) = 1$, and $p \circ f(\infty) = \infty$.

Since $p \circ f$ takes circles in $\overline{\mathbb{C}}$ to circles in $\overline{\mathbb{C}}$, it must be that $p \circ f(\mathbb{R}) = \mathbb{R}$, since $p \circ f(\infty) = \infty$ and $\overline{\mathbb{R}}$ is the circle in $\overline{\mathbb{C}}$ determined by the triple $(0, 1, \infty)$.

Since $p \circ f$ fixes ∞ and takes $\mathbb{R}$ to $\mathbb{R}$, either $p \circ f(\mathbb{H}) = \mathbb{H}$ or $p \circ f(\mathbb{H})$ is the lower half-plane. In the former case, set $m = p$. In the latter case, set $m = C \circ p$, where $C(z) = \overline{z}$ is complex conjugation.

We now have an element m of Möb so that $m \circ f(0) = 0$, $m \circ f(1) = 1$, $m \circ f(\infty) = \infty$, and $m \circ f(\mathbb{H}) = \mathbb{H}$. We show that $m \circ f$ is the identity. We do this by constructing a dense set of points in $\overline{\mathbb{C}}$, each of which is fixed by $m \circ f$.

Since $m \circ f$ fixes ∞ and lies in $\text{Homeo}^{\text{C}}(\overline{\mathbb{C}})$, we see that $m \circ f$ takes Euclidean lines in $\mathbb{C}$ to Euclidean lines in $\mathbb{C}$, and takes Euclidean circles in $\mathbb{C}$ to Euclidean circles in $\mathbb{C}$.

Before beginning the construction of this dense set, we introduce a bit of notation. Set $Z = \{z \in \overline{\mathbb{C}} : m \circ f(z) = z\}$ to be the set of points of $\overline{\mathbb{C}}$ fixed by $m \circ f$. By our choice of m, we have that 0, 1, and ∞ are points of Z.

Also, if X and Y are two Euclidean lines in $\mathbb{C}$ that intersect at some point z_0, and if $m \circ f(X) = X$ and $m \circ f(Y) = Y$, then $m \circ f(z_0) = z_0$ and so z_0 is contained in this set Z of points fixed by $m \circ f$.

For each $s \in \mathbb{R}$, let $V(s)$ be the vertical line in $\mathbb{C}$ through s and let $H(s)$ be the horizontal line in $\mathbb{C}$ through is.

Let H be any horizontal line in $\mathbb{C}$. Since $m \circ f(\mathbb{R}) = \mathbb{R}$ and since H and $\mathbb{R}$ are disjoint, we see that $m \circ f(H)$ and $m \circ f(\mathbb{R}) = \mathbb{R}$ are disjoint lines in $\mathbb{C}$, and so H is again a horizontal line in $\mathbb{C}$. Also, since $m \circ f(\mathbb{H}) = \mathbb{H}$, we have that H lies in $\mathbb{H}$ if and only if $m \circ f(H)$ lies in $\mathbb{H}$.

Let A be the Euclidean circle with Euclidean centre $\frac{1}{2}$ and Euclidean radius $\frac{1}{2}$. Since $V(0)$ is tangent to A at 0, we see that $m \circ f(V(0))$ is the tangent line to $m \circ f(A)$ at $m \circ f(0) = 0$, and similarly that $m \circ f(V(1))$ is the tangent line to $m \circ f(A)$ at 1.

Since $V(0)$ and $V(1)$ are parallel Euclidean lines in $\mathbb{C}$, we see that $m \circ f(V(0))$ and $m \circ f(V(1))$ are also parallel Euclidean lines in $\mathbb{C}$, and so we must have that $m \circ f(V(0)) = V(0)$ and $m \circ f(V(1)) = V(1)$.

In particular, this forces $m \circ f(A) = A$, as the tangent lines through 0 and 1 to any other Euclidean circle passing through 0 and 1 are not parallel. However, even though $m \circ f(A) = A$, we do not yet know that $A \cap Z$ contains any points other than 0 and 1.

However, we can run the same argument with the two horizontal tangent lines to A. Consider first the tangent line $H(\frac{1}{2})$ to A at $\frac{1}{2} + i\frac{1}{2}$. Since $m \circ f(H(\frac{1}{2}))$ is again a horizontal line in $\mathbb{H}$ tangent to $m \circ f(A) = A$, we see that $m \circ f(H(\frac{1}{2})) = H(\frac{1}{2})$.

We now have more points in Z. Namely, the intersections $H(\frac{1}{2}) \cap V(0) = i\frac{1}{2}$ and $H(\frac{1}{2}) \cap V(1) = 1 + i\frac{1}{2}$ lie in Z. The same argument gives that $m \circ f(H(-\frac{1}{2})) = H(-\frac{1}{2})$, and hence that both $H(-\frac{1}{2}) \cap V(0) = -\frac{1}{2}i$ and $H(-\frac{1}{2}) \cap V(1) = 1 - \frac{1}{2}i$ lie in Z.

Each pair of points in Z gives rise to a Euclidean line that is taken to itself by $m \circ f$, and each triple of points in Z gives rise to a Euclidean circle that is taken to itself by $m \circ f$. The intersections of these Euclidean lines and Euclidean circles give rise to more points of Z, which in turn give rise to more Euclidean lines and Euclidean circles taken to themselves, and so on.

Continuing on, this process yields that Z contains a dense set of points of $\overline{\mathbb{C}}$, which in turn implies that $m \circ f$ is the identity, by Exercise 1.16. Hence,

$f = m^{-1}$ is an element of Möb. This completes the sketch of the proof of Theorem 2.21.

2.7 The Conformality of Elements of Möb

In this section, we describe the last major property of Möb that we will make use of. We begin with a definition.

Definition 2.22

Given two smooth curves C_1 and C_2 in $\mathbb{C}$ that intersect at a point z_0, define the *angle* $\text{angle}(C_1, C_2)$ *between* C_1 *and* C_2 at z_0 to be the angle between the tangent lines to C_1 and C_2 at z_0, measured from C_1 to C_2.

In our measurement of angle, we adopt the convention that counterclockwise angles are positive and clockwise angles are negative. By this definition of angle, we have that

$$\text{angle}(C_2, C_1) = -\text{angle}(C_1, C_2).$$

Note that angle as we have defined it is not a well-defined notion, but instead is defined only up to additive multiples of π. If we were to be formal, we would really need to define angle to take its values in $\mathbb{R}/\pi\mathbb{Z}$. However, this ambiguity in the definition of angle causes us no difficulty in this section.

A homeomorphism of $\overline{\mathbb{C}}$ that preserves the absolute value of the angle between curves is said to be *conformal.* We note that this usage is slightly non-standard, as many authors use conformal to mean that the actual angles, and not merely the absolute values of the angles, are preserved.

The last major fact about Möb we need to establish is that the elements of Möb are conformal. The proof we give here is analytic. Though we do not give it here, it is possible to give a geometric proof using stereographic projection. See for example Jones and Singerman [14].

Theorem 2.23

The elements of Möb are conformal homeomorphisms of $\overline{\mathbb{C}}$.

The proof of Theorem 2.23 contains a number of calculations left to the reader.

Since the angle between two curves is by definition the angle between their tangent lines, it suffices to check whether the angle angle(X_1, X_2) between X_1 and X_2 is equal to the angle angle$(m(X_1), m(X_2))$ between $m(X_1)$ and $m(X_2)$, where X_1 and X_2 are Euclidean lines in $\mathbb{C}$.

So, let X_1 and X_2 be two Euclidean lines in $\mathbb{C}$ that intersect at a point z_0, let z_k be a point on X_k not equal to z_0, and let s_k be the slope of X_k. These quantities are connected by the formula

$$s_k = \frac{\operatorname{Im}(z_k - z_0)}{\operatorname{Re}(z_k - z_0)}.$$

Let θ_k be the angle that X_k makes with the real axis $\mathbb{R}$, and note that

$$s_k = \tan(\theta_k).$$

In particular, the angle angle(X_1, X_2) between X_1 and X_2 is given by

$$\operatorname{angle}(X_1, X_2) = \theta_2 - \theta_1 = \arctan(s_2) - \arctan(s_1).$$

We know from Section 2.6 that Möb is generated by the transformations of the form $f(z) = az + b$ for a, $b \in \mathbb{C}$ and $a \neq 0$, as well as the two transformations $J(z) = \frac{1}{z}$ and $C(z) = \overline{z}$. We take each in turn.

Consider $f(z) = az + b$, where a, $b \in \mathbb{C}$ and $a \neq 0$. Write $a = \rho e^{i\beta}$. Since $f(\infty) = \infty$, both $f(X_1)$ and $f(X_1)$ are again Euclidean lines in $\mathbb{C}$. Since $f(X_k)$ passes through the points $f(z_0)$ and $f(z_k)$, the slope t_k of the Euclidean line $f(X_k)$ is

$$\begin{aligned} t_k = \frac{\operatorname{Im}(f(z_k) - f(z_0))}{\operatorname{Re}(f(z_k) - f(z_0))} &= \frac{\operatorname{Im}(a(z_k - z_0))}{\operatorname{Re}(a(z_k - z_0))} \\ &= \frac{\operatorname{Im}(e^{i\beta}(z_k - z_0))}{\operatorname{Re}(e^{i\beta}(z_k - z_0))} = \tan(\beta + \theta_k). \end{aligned}$$

In particular, we see that

$$\begin{aligned} \operatorname{angle}(f(X_1), f(X_2)) &= \arctan(t_2) - \arctan(t_1) \\ &= (\beta + \theta_2) - (\beta + \theta_1) \\ &= \theta_2 - \theta_1 = \operatorname{angle}(X_1, X_2), \end{aligned}$$

and so m is conformal.

Consider now $J(z) = \frac{1}{z}$. Here, we need to take a slightly different approach, since $J(X_1)$ and $J(X_2)$ no longer need be Euclidean lines in $\mathbb{C}$, but instead may both be Euclidean circles in $\mathbb{C}$ that intersect at 0, or one might be a Euclidean line and the other a Euclidean circle. We work through the case that both are Euclidean circles, and leave the other cases to the reader.

So, we may suppose that X_k is given as the set of solutions to the equation

$$\beta_k z + \overline{\beta_k} \overline{z} + 1 = 0,$$

where $\beta_k \in \mathbb{C}$. The slope of X_k is then given by

$$s_k = \frac{\mathrm{Re}(\beta_k)}{\mathrm{Im}(\beta_k)}.$$

Given the form of the equation for X_k, we also know that $J(X_k)$ is the set of solutions to the equation

$$z\overline{z} + \overline{\beta_k} z + \beta_k \overline{z} = 0,$$

which we can rewrite as

$$|z + \beta_k|^2 = |\beta_k|^2,$$

so that $J(X_k)$ is the Euclidean circle with Euclidean centre $-\beta_k$ and Euclidean radius $|\beta_k|$.

The slope of the tangent line to $J(X_k)$ at 0 is then

$$-\frac{\mathrm{Re}(\beta_k)}{\mathrm{Im}(\beta_k)} = -\tan(\theta_k) = \tan(-\theta_k),$$

and so $J(X_k)$ makes angle $-\theta_k$ with $\mathbb{R}$.

The angle between $J(X_1)$ and $J(X_2)$ is then given by

$$\mathrm{angle}(J(X_1), J(X_2)) = -\theta_2 - (-\theta_1) = -\mathrm{angle}(X_1, X_2),$$

and so J is conformal.

Exercise 2.34

Show that $C(z) = \overline{z}$ is conformal.

This completes the proof of Theorem 2.23.

Examining the proof carefully, we see that each $f(z) = az + b$ preserves the sign of the angle between X_1 and X_2 as well, while $C(z) = \overline{z}$ reverses the sign of the angle.

There is a subtlety with regard to $J(z) = \frac{1}{z}$. The angle between $J(X_1)$ and $J(X_2)$ at 0 is the angle between X_1 and X_2 at ∞, which is the negative of the angle between X_1 and X_2 at z_0. Hence, J also preserves the sign of the angle between X_1 and X_2.

Hence, every element of $\mathrm{Möb}^+$ preserves the sign of the angle between X_1 and X_2, since $\mathrm{Möb}^+$ is generated by $J(z) = \frac{1}{z}$ and the $f(z) = az + b$ for $a, b \in \mathbb{C}$ with $a \neq 0$.

2.8 Preserving $\mathbb{H}$

Recall that our foray into Möbius transformations and the general Möbius group was undertaken as an attempt to determine those transformations of the upper half-plane $\mathbb{H}$ that take hyperbolic lines to hyperbolic lines.

One place to look for such transformations is the subgroup of Möb preserving $\mathbb{H}$. So, consider the group

$$\text{Möb}(\mathbb{H}) = \{m \in \text{Möb} \mid m(\mathbb{H}) = \mathbb{H}\}.$$

Theorem 2.24

Every element of Möb($\mathbb{H}$) takes hyperbolic lines in $\mathbb{H}$ to hyperbolic lines in $\mathbb{H}$.

The proof of Theorem 2.24 is an immediate consequence of Theorem 2.23, which states that the elements of Möb($\mathbb{H}$) preserve angles between circles in $\overline{\mathbb{C}}$, together with the fact that every hyperbolic line in $\mathbb{H}$ is the intersection of $\mathbb{H}$ with a circle in $\overline{\mathbb{C}}$ perpendicular to $\overline{\mathbb{R}}$.

Let

$$\text{Möb}^+(\mathbb{H}) = \{m \in \text{Möb}^+ \mid m(\mathbb{H}) = \mathbb{H}\}$$

be the subgroup of Möb($\mathbb{H}$) consisting of the Möbius transformations preserving the upper half-plane $\mathbb{H}$.

These definitions are somehow unsatisfying, as we do not have an explicit expression for the element of either Möb($\mathbb{H}$) or Möb$^+$($\mathbb{H}$). We spend the remainder of this section deriving these desired explicit expressions.

Since $\mathbb{H}$ is a disc in $\overline{\mathbb{C}}$ determined by the circle $\overline{\mathbb{R}}$ in $\overline{\mathbb{C}}$, we first determine the explicit form of an element of the subgroup

$$\text{Möb}(\overline{\mathbb{R}}) = \{m \in \text{Möb} \mid m(\overline{\mathbb{R}}) = \overline{\mathbb{R}}\}.$$

We know from Exercise 2.31 that every element of Möb can be written either as $m(z) = \frac{az+b}{cz+d}$ or as $m(z) = \frac{a\overline{z}+b}{c\overline{z}+d}$, where a, b, c, $d \in \mathbb{C}$ and $ad - bc = 1$. We are interested in determining the conditions imposed on a, b, c, and d by assuming that $m(\overline{\mathbb{R}}) = \overline{\mathbb{R}}$.

Note that in the latter case, we may consider instead the composition

$$m \circ C(z) = m(\overline{z}) = \frac{az+b}{cz+d},$$

and so reduce ourselves to considering just the former case that

$$m(z) = \frac{az+b}{cz+d},$$

where a, b, c, $d \in \mathbb{C}$ and $ad - bc = 1$.

Since m takes $\overline{\mathbb{R}}$ to $\overline{\mathbb{R}}$, we have that the three points

$$m^{-1}(\infty) = -\frac{d}{c}, \ m(\infty) = \frac{a}{c} \text{ and } m^{-1}(0) = -\frac{b}{a}$$

all lie in $\overline{\mathbb{R}}$.

Suppose for the moment that $a \neq 0$ and $c \neq 0$, so that these three points all lie in $\mathbb{R}$. We can then express each coefficient of m as a multiple of c. Specifically, we have that $a = m(\infty)c$, $b = -m^{-1}(0)a = -m^{-1}(0)m(\infty)c$, and $d = -m^{-1}(\infty)c$. In particular, we can rewrite m as

$$m(z) = \frac{az+b}{cz+d} = \frac{m(\infty)cz - m^{-1}(0)m(\infty)c}{cz - m^{-1}(\infty)c}.$$

However, normalizing so that the determinant of m is 1 imposes a condition on c, namely that

$$\begin{aligned} 1 = ad - bc &= c^2 \left[-m(\infty)m^{-1}(\infty) + m(\infty)m^{-1}(0)\right] \\ &= c^2 \left[m(\infty)(m^{-1}(0) - m^{-1}(\infty))\right]. \end{aligned}$$

Since $m(\infty)$, $m^{-1}(0)$, and $m^{-1}(\infty)$ are all real, this implies that c is either real or purely imaginary, and hence that the coefficients of m are either all real or all purely imaginary.

Exercise 2.35

Complete this analysis of the coefficients of m by considering the two remaining cases, namely that $a = 0$ and that $c = 0$.

Conversely, if m has either the form $m(z) = \frac{az+b}{cz+d}$ with $ad - bc = 1$, or the form $m(z) = \frac{a\overline{z}+b}{c\overline{z}+d}$ with $ad - bc = 1$, where the coefficients of m are either all real or all purely imaginary, then the three points $m(0)$, $m(\infty)$, and $m^{-1}(\infty)$ all lie in $\overline{\mathbb{R}}$, and so m takes $\overline{\mathbb{R}}$ to $\overline{\mathbb{R}}$.

We summarize this analysis in the following theorem.

Theorem 2.25

Every element of $\text{Möb}(\overline{\mathbb{R}})$ has one of the following four forms:

1 $m(z) = \frac{az+b}{cz+d}$ with $a, b, c, d \in \mathbb{R}$ and $ad - bc = 1$;

2 $m(z) = \frac{a\overline{z}+b}{c\overline{z}+d}$ with $a, b, c, d \in \mathbb{R}$ and $ad - bc = 1$;

3 $m(z) = \frac{az+b}{cz+d}$ with a, b, c, d purely imaginary and $ad - bc = 1$;

4 $m(z) = \frac{a\overline{z}+b}{c\overline{z}+d}$ with a, b, c, d purely imaginary and $ad - bc = 1$.

Note that we now also have an explicit form for the subgroup

$$\text{Möb}(A) = \{m \in \text{Möb} \mid m(A) = A\}$$

of Möb for any circle A in $\overline{\mathbb{C}}$. All we need do is choose some element p of Möb satisfying $p(\overline{\mathbb{R}}) = A$, and consider

$$\{p \circ m \circ p^{-1} \mid m \in \text{Möb}(\overline{\mathbb{R}})\}.$$

If n is any element of Möb satisfying $n(A) = A$, then $p^{-1} \circ n \circ p(\overline{\mathbb{R}}) = \overline{\mathbb{R}}$. Hence, we can write $p^{-1} \circ n \circ p = m$ for some element m of Möb$(\overline{\mathbb{R}})$, and so $n = p \circ m \circ p^{-1}$. This gives that

$$\text{Möb}(A) = \{p \circ m \circ p^{-1} \mid m \in \text{Möb}(\overline{\mathbb{R}})\}.$$

This subgroup Möb(A) is independent of the choice of p. To see this, suppose that q is another element of Möb satisfying $q(\overline{\mathbb{R}}) = A$. Then, $p^{-1} \circ q$ takes $\overline{\mathbb{R}}$ to $\overline{\mathbb{R}}$, and so we can write $q = p \circ t$ for some element t of Möb$(\overline{\mathbb{R}})$.

Hence, for any m in Möb$(\overline{\mathbb{R}})$ we have that $q \circ m \circ q^{-1} = p \circ (t \circ m \circ t^{-1}) \circ p^{-1}$, and so

$$\{p \circ m \circ p^{-1} \mid m \in \text{Möb}(\overline{\mathbb{R}})\} = \{q \circ m \circ q^{-1} \mid m \in \text{Möb}(\overline{\mathbb{R}})\}.$$

Exercise 2.36

Determine the general form of an element of Möb$(\mathbb{S}^1)$.

We are now ready to determine Möb$(\mathbb{H})$. Each element of Möb$(\overline{\mathbb{R}})$ either preserves each of the two discs in $\overline{\mathbb{C}}$ determined by $\overline{\mathbb{R}}$, namely the upper and lower half-planes, or interchanges them. In order to determine which, we consider the image of a single point in one of the discs.

Specifically, an element m of Möb$(\overline{\mathbb{R}})$ is an element of Möb$(\mathbb{H})$ if and only if the imaginary part of $m(i)$ is positive. So, we need to check the value of $\text{Im}(m(i))$ for each of the four possible forms of an element of Möb$(\overline{\mathbb{R}})$.

If m has the form $m(z) = \frac{az+b}{cz+d}$, where a, b, c, and d are real and $ad - bc = 1$, then the imaginary part of $m(i)$ is given by

$$\begin{aligned}
\mathrm{Im}(m(i)) &= \mathrm{Im}\left(\frac{ai+b}{ci+d}\right) \\
&= \mathrm{Im}\left(\frac{(ai+b)(-ci+d)}{(ci+d)(-ci+d)}\right) = \frac{ad-bc}{c^2+d^2} = \frac{1}{c^2+d^2} > 0,
\end{aligned}$$

and so m lies in $\mathrm{M\ddot{o}b}(\mathbb{H})$.

If m has the form $m(z) = \frac{a\overline{z}+b}{c\overline{z}+d}$, where a, b, c, and d are real and $ad - bc = 1$, then the imaginary part of $m(i)$ is given by

$$\begin{aligned}
\mathrm{Im}(m(i)) &= \mathrm{Im}\left(\frac{-ai+b}{-ci+d}\right) \\
&= \mathrm{Im}\left(\frac{(-ai+b)(ci+d)}{(-ci+d)(ci+d)}\right) = \frac{-ad+bc}{c^2+d^2} = \frac{-1}{c^2+d^2} < 0,
\end{aligned}$$

and so m does not lie in $\mathrm{M\ddot{o}b}(\mathbb{H})$.

If m has the form $m(z) = \frac{az+b}{cz+d}$, where a, b, c, and d are purely imaginary and $ad - bc = 1$, write $a = \alpha i$, $b = \beta i$, $c = \gamma i$, and $d = \delta i$, so that $\alpha\delta - \beta\gamma = -1$. Then, the imaginary part of $m(i)$ is given by

$$\begin{aligned}
\mathrm{Im}(m(i)) &= \mathrm{Im}\left(\frac{ai+b}{ci+d}\right) = \mathrm{Im}\left(\frac{-\alpha+\beta i}{-\gamma+\delta i}\right) \\
&= \mathrm{Im}\left(\frac{(-\alpha+\beta i)(-\gamma-\delta i)}{(-\gamma+\delta i)(-\gamma-\delta i)}\right) = \frac{\alpha\delta-\beta\gamma}{\gamma^2+\delta^2} = \frac{-1}{\gamma^2+\delta^2} < 0,
\end{aligned}$$

and so m does not lie in $\mathrm{M\ddot{o}b}(\mathbb{H})$.

If m has the form $m(z) = \frac{a\overline{z}+b}{c\overline{z}+d}$, where a, b, c, and d are purely imaginary and $ad - bc = 1$, write $a = \alpha i$, $b = \beta i$, $c = \gamma i$, and $d = \delta i$, so that $\alpha\delta - \beta\gamma = -1$. Then, the imaginary part of $m(i)$ is given by

$$\begin{aligned}
\mathrm{Im}(m(i)) &= \mathrm{Im}\left(\frac{-ai+b}{-ci+d}\right) = \mathrm{Im}\left(\frac{\alpha+\beta i}{\gamma+\delta i}\right) \\
&= \mathrm{Im}\left(\frac{(\alpha+\beta i)(\gamma-\delta i)}{(\gamma+\delta i)(\gamma-\delta i)}\right) = \frac{-\alpha\delta+\beta\gamma}{\gamma^2+\delta^2} = \frac{1}{\gamma^2+\delta^2} > 0,
\end{aligned}$$

and so m lies in $\mathrm{M\ddot{o}b}(\mathbb{H})$.

We summarize this analysis in the following theorem.

Theorem 2.26

Every element of $\mathrm{M\ddot{o}b}(\mathbb{H})$ either has the form

$$m(z) = \frac{az+b}{cz+d}, \text{ where } a,\ b,\ c,\ d \in \mathbb{R} \text{ and } ad - bc = 1,$$

or has the form

$$n(z) = \frac{a\overline{z} + b}{c\overline{z} + d}, \text{ where } a,\ b,\ c,\ d \text{ are purely imaginary and } ad - bc = 1.$$

One consequence of Theorem 2.26 is that every element of $\text{Möb}^+(\mathbb{H})$ has the form

$$m(z) = \frac{az + b}{cz + d}, \text{ where } a,\ b,\ c,\ d \in \mathbb{R} \text{ and } ad - bc = 1,$$

since no element of $\text{Möb}(\mathbb{H})$ of the form

$$n(z) = \frac{a\overline{z} + b}{c\overline{z} + d}, \text{ where } a,\ b,\ c,\ d \text{ are purely imaginary and } ad - bc = 1$$

can be an element of $\text{Möb}^+(\mathbb{H})$.

Exercise 2.37

Show that $\text{Möb}(\mathbb{H})$ is generated by elements of the form $m(z) = az + b$ for $a > 0$ and $b \in \mathbb{R}$, $K(z) = \frac{-1}{z}$, and $B(z) = -\overline{z}$.

Exercise 2.38

Write down the general form of an element of $\text{Möb}(\mathbb{D})$, where $\mathbb{D} = \{z \in \mathbb{C} : |z| < 1\}$ is the unit disc in $\mathbb{C}$.

Note that we have not actually addressed the question of whether $\text{Möb}(\mathbb{H})$ contains all the transformations of $\mathbb{H}$ that take hyperbolic lines to hyperbolic lines. We have merely shown that every element of $\text{Möb}(\mathbb{H})$ has this property, which will suffice for the time being.

We close this section by showing that this characterization of the general form of an element of $\text{Möb}(\overline{\mathbb{R}})$ is exactly what we need in order to show that the definition of reflection in a circle in $\overline{\mathbb{C}}$ given in Section 2.6 is well-defined.

Proposition 2.27

Reflection in a circle in $\overline{\mathbb{C}}$, as defined in Section 2.6, is well-defined.

For any element m of $\text{Möb}(\overline{\mathbb{R}})$, a direct calculation based on the two possible forms for m shows that $C \circ m = m \circ C$, where $C(z) = \overline{z}$ is complex conjugation. If $m(z) = \frac{az+b}{cz+d}$ where $a,\ b,\ c,\ d \in \mathbb{R}$ and $ad - bc = 1$, then

$$C \circ m(z) = \frac{a\overline{z} + b}{c\overline{z} + d} = m \circ C(z).$$

If $m(z) = \frac{a\overline{z}+b}{c\overline{z}+d}$ where a, b, c, d are purely imaginary and $ad - bc = 1$, then

$$C \circ m(z) = \frac{-az - b}{-cz - d} = \frac{az + b}{cz + d} = m \circ C(z).$$

Let A be a circle in $\overline{\mathbb{C}}$ and let m and n be two elements of Möb($\overline{\mathbb{R}}$) taking $\overline{\mathbb{R}}$ to A. Then, $n^{-1} \circ m$ takes $\overline{\mathbb{R}}$ to $\overline{\mathbb{R}}$, and so $n^{-1} \circ m = p$ for some element p of Möb($\overline{\mathbb{R}}$). In particular, $p \circ C = C \circ p$. Write $m = n \circ p$, and calculate that

$$m \circ C \circ m^{-1} = n \circ p \circ C \circ p^{-1} \circ n^{-1} = n \circ p \circ p^{-1} \circ C \circ n^{-1} = n \circ C \circ n^{-1}.$$

Hence, reflection in a circle in $\overline{\mathbb{C}}$ is well-defined.

2.9 Transitivity Properties of Möb($\mathbb{H}$)

In Section 2.2, we described some sets on which Möb acts transitively, and we have also seen some ways in which knowing the transitivity of the action of Möb$^+$ on these sets is useful. In this section, we restrict our attention to the action of Möb($\mathbb{H}$) on $\mathbb{H}$, and show that we can obtain similar sorts of results.

We first observe that Möb($\mathbb{H}$) acts transitively on $\mathbb{H}$ itself. That is, for each pair w_1 and w_2 of distinct points of $\mathbb{H}$, there exists an element m in Möb($\mathbb{H}$) taking w_1 to w_2. Even though we know that Möb acts transitively on triples of distinct points of $\overline{\mathbb{C}}$, it is not *a priori* obvious that there exists an element of Möb that both takes $\mathbb{H}$ to itself and takes w_1 to w_2.

Proposition 2.28

Möb($\mathbb{H}$) acts transitively on $\mathbb{H}$.

Using Lemma 2.8, it suffices to show that for any point w of $\mathbb{H}$, there exists an element m of Möb($\mathbb{H}$) satisfying $m(w) = i$.

Write $w = a+ib$, where $a, b \in \mathbb{R}$ and $b > 0$. We construct an element of Möb($\mathbb{H}$) taking w to i as a composition. We first move w to the positive imaginary axis using $p(z) = z - a$, so that $p(w) = p(a + ib) = bi$.

We next apply $q(z) = \frac{1}{b}z$ to $p(w)$, so that $q(p(w)) = q(bi) = i$. Note that since $-a \in \mathbb{R}$ and $\frac{1}{b} > 0$, we have by Theorem 2.26 that both $p(z)$ and $q(z)$, and hence $q \circ p(z)$, lie in Möb($\mathbb{H}$). This completes the proof of Proposition 2.28.

Exercise 2.39

Show that Möb($\mathbb{H}$) acts transitively on the set $\mathcal{L}$ of hyperbolic lines in $\mathbb{H}$.

Exercise 2.40

Give an explicit expression for an element of Möb($\mathbb{H}$) taking the hyperbolic line ℓ determined by 1 and -2 to the positive imaginary axis I.

Even though Möb($\mathbb{H}$) acts transitively on the set $\mathcal{L}$ of hyperbolic lines in $\mathbb{H}$ and even though a hyperbolic line is determined by a pair of distinct points in $\mathbb{H}$, it does not follow that Möb($\mathbb{H}$) acts transitively on the set $\mathcal{P}$ of pairs of distinct points of $\mathbb{H}$, much less on the set $\mathcal{T}_{\mathbb{H}}$ of triples of distinct points of $\mathbb{H}$.

We can see this directly by considering the positive imaginary axis I. Since the endpoints at infinity of I are 0 and ∞, every element of Möb($\mathbb{H}$) taking I to itself either fixes both 0 and ∞, or else interchanges them. Recall from Theorem 2.26 that we know the general form of an element of Möb($\mathbb{H}$). Namely, an element m of Möb($\mathbb{H}$) fixing both 0 and ∞ either has the form $m(z) = az$, where $a \in \mathbb{R}$ and $a > 0$, or has the form $m(z) = -a\overline{z}$, where again $a \in \mathbb{R}$ and $a > 0$.

An element m of Möb($\mathbb{H}$) interchanging 0 and ∞ either has the form $m(z) = -\frac{b}{z}$, where $b \in \mathbb{R}$ and $b > 0$, or has the form $m(z) = \frac{b}{\overline{z}}$, where again $b \in \mathbb{R}$ and $b > 0$.

In any of these cases, we can see that there is no element of Möb($\mathbb{H}$) that takes the positive imaginary axis I to itself, that takes i to i, and that takes $2i$ to $3i$. In fact, the only element of Möb($\mathbb{H}$), other than the identity, that takes I to itself and fixes i is $B(z) = -\overline{z}$, that is reflection in I and hence fixes every point of I.

We will return to this failure of Möb($\mathbb{H}$) to act transitively on the set $\mathcal{P}$ of pairs of distinct points of $\mathbb{H}$ after we have developed a means of measuring hyperbolic distance in $\mathbb{H}$.

We also need to make use of the analog in $\mathbb{H}$ of a disc in $\overline{\mathbb{C}}$.

Definition 2.29

A *half-plane in* $\mathbb{H}$ is a component of the complement of a hyperbolic line in $\mathbb{H}$.

In particular, each half-plane is determined by a unique hyperbolic line, and each hyperbolic line determines a pair of half-planes.

The hyperbolic line determining a half-plane is the *bounding line* for the half-plane. A half-plane is *closed* if it is the union of a hyperbolic line ℓ with one of the components of $\mathbb{H} - \ell$, and is *open* if it is just one of the components of $\mathbb{H} - \ell$.

In much the same way that we extended the transitivity of Möb on the set $\mathcal{C}$ of circles in $\overline{\mathbb{C}}$ to transitivity on the set $\mathcal{D}$ of discs in $\overline{\mathbb{C}}$, we can extend the transitivity of Möb($\mathbb{H}$) on the set $\mathcal{L}$ of hyperbolic lines in $\mathbb{H}$ to transitivity on the set $\mathcal{H}$ of half-planes in $\mathbb{H}$.

Exercise 2.41

Show that Möb($\mathbb{H}$) acts transitively on the set $\mathcal{H}$ of open half-planes in $\mathbb{H}$.

We can also consider the action of Möb($\mathbb{H}$) on the boundary at infinity $\overline{\mathbb{R}}$ of $\mathbb{H}$.

Proposition 2.30

Möb($\mathbb{H}$) acts triply transitively on the set $\mathcal{T}_{\overline{\mathbb{R}}}$ of triples of distinct points of $\overline{\mathbb{R}}$.

Again using Lemma 2.8, given a triple (z_1, z_2, z_3) of distinct points of $\overline{\mathbb{R}}$, it suffices to show that there exists an element of Möb($\mathbb{H}$) taking (z_1, z_2, z_3) to $(0, 1, \infty)$.

Let ℓ be the hyperbolic line whose endpoints at infinity are z_1 and z_3, and let m be an element of Möb($\mathbb{H}$) taking ℓ to the positive imaginary axis I. By composing m with $K(z) = -\frac{1}{z}$ if necessary, we can assume that $m(z_1) = 0$ and $m(z_3) = \infty$. Set $b = m(z_2)$.

If $b > 0$, then the composition of m with $p(z) = \frac{1}{b}z$ takes (z_1, z_2, z_3) to $(0, 1, \infty)$.

If $b < 0$, then $p(z) = \frac{1}{b}z$ no longer lies in Möb($\mathbb{H}$), but the composition of m with $q(z) = \frac{1}{b}\overline{z}$, which does lie in Möb($\mathbb{H}$), takes (z_1, z_2, z_3) to $(0, 1, \infty)$. This completes the proof of Proposition 2.30.

We close this section by noting that Möb$^+$($\mathbb{H}$) does not act triply transitively on $\mathcal{T}_{\overline{\mathbb{R}}}$, since there is no element of Möb$^+$($\mathbb{H}$) taking $(0, 1, \infty)$ to $(0, -1, \infty)$.

3
Length and Distance in $\mathbb{H}$

We now have a reasonable group of transformations of $\mathbb{H}$, namely Möb($\mathbb{H}$). This group is reasonable in the sense that its elements take hyperbolic lines to hyperbolic lines and preserve angles. In this chapter, we derive a means of measuring lengths of paths in $\mathbb{H}$ which is invariant under the action of this group, expressed as an *invariant element of arc-length*. From this invariant element of arc-length, we construct an *invariant notion of distance* on $\mathbb{H}$ and explore some of its basic properties.

3.1 Paths and Elements of Arc-length

Now that we have a group of transformations of $\mathbb{H}$ taking hyperbolic lines to hyperbolic lines, namely Möb($\mathbb{H}$), we are in a position to attempt to derive the element of arc-length for the hyperbolic metric on $\mathbb{H}$. However, we first need to recall from calculus the definition of an element of arc-length.

A *path* in the plane $\mathbb{R}^2$ is a differentiable function $f : [a,b] \to \mathbb{R}^2$, given by $f(t) = (x(t), y(t))$, where $x(t)$ and $y(t)$ are differentiable functions of t and where $[a,b]$ is some interval in $\mathbb{R}$. The image of an interval $[a,b]$ under a path f is a *curve* in $\mathbb{R}^2$.

The *Euclidean length* of f is given by the integral

$$\text{length}(f) = \int_a^b \sqrt{(x'(t))^2 + (y'(t))^2}\, \mathrm{d}t,$$

where $\sqrt{(x'(t))^2 + (y'(t))^2}\, \mathrm{d}t$ is the *element of arc-length* in $\mathbb{R}^2$.

Note that the length of a graph of a differentiable function $g : [a, b] \to \mathbb{R}$ is a special case of the length of a path as described above. In this case, given g we construct a path $f : [a, b] \to \mathbb{R}^2$ by setting $f(t) = (t, g(t))$.

As an example, consider the path $f : [0, 2] \to \mathbb{R}^2$ given by $f(t) = (1 + t, \frac{1}{2}t^2)$. The length of f is

$$\begin{aligned} \text{length}(f) = \int_0^2 \sqrt{1+t^2}\, \mathrm{d}t &= \frac{1}{2}[t\sqrt{1+t^2} + \ln|t + \sqrt{1+t^2}|]\,\Big|_0^2 \\ &= \sqrt{5} + \frac{1}{2}\ln(2 + \sqrt{5}). \end{aligned}$$

We now engage in a bit of notational massage. If we view f as a path into $\mathbb{C}$ instead of $\mathbb{R}^2$ and write $f(t) = x(t) + y(t)i$, we then have that $f'(t) = x'(t) + y'(t)i$ and $|f'(t)| = \sqrt{(x'(t))^2 + (y'(t))^2}$. In particular, the integral for the length of f becomes

$$\text{length}(f) = \int_a^b \sqrt{(x'(t))^2 + (y'(t))^2}\, \mathrm{d}t = \int_a^b |f'(t)|\, \mathrm{d}t,$$

and so we can write the standard element of arc-length in $\mathbb{C}$ as

$$|\mathrm{d}z| = |f'(t)|\, \mathrm{d}t.$$

At this point, we introduce a new piece of notation and abbreviate the integral on the right hand side as

$$\int_a^b |f'(t)|\, \mathrm{d}t = \int_f |\mathrm{d}z|.$$

One advantage to this notation is that it is extremely flexible and easily extendable. For instance, we may easily write any path integral in this notation. That is, let ρ be a continuous function $\rho : \mathbb{C} \to \mathbb{R}$. The *path integral* of ρ along a path $f : [a, b] \to \mathbb{C}$ is given by the integral

$$\int_f \rho(z)\, |\mathrm{d}z| = \int_a^b \rho(f(t))\, |f'(t)|\, \mathrm{d}t.$$

We can interpret this path integral as given a new element of arc-length, denoted $\rho(z)\,|\mathrm{d}z|$, given by scaling the Euclidean element of arc-length $|\mathrm{d}z|$ at every point, where the amount of scaling is given by the function ρ. This gives rise to the following definition.

Definition 3.1

For a differentiable path $f : [a, b] \to \mathbb{C}$, we define the *length of f with respect to the element of arc-length $\rho(z)|\mathrm{d}z|$* to be the path integral

$$\text{length}_\rho(f) = \int_f \rho(z)\,|\mathrm{d}z| = \int_a^b \rho(f(t))\,|f'(t)|\,\mathrm{d}t.$$

There are innumerable variations on this theme, and we spend the remainder of this section exploring some of them. In the next section, we restrict consideration to such elements of arc-length on $\mathbb{H}$.

As a specific example, set $\rho(z) = \frac{1}{1+|z|^2}$ and consider the element of arc-length

$$\rho(z)\,|\mathrm{d}z| = \frac{1}{1+|z|^2}|\mathrm{d}z|$$

on $\mathbb{C}$.

For $r > 0$, consider the path $f : [0, 2\pi] \to \mathbb{C}$ given by $f(t) = re^{it}$, which parametrizes the Euclidean circle with Euclidean centre 0 and Euclidean radius r. Since $|f(t)| = r$ and $|f'(t)| = |ire^{it}| = r$, the length of f with respect to the element of arc-length $\frac{1}{1+|z|^2}\,|\mathrm{d}z|$ is

$$\text{length}_\rho(f) = \int_f \frac{1}{1+|z|^2}|\mathrm{d}z| = \int_0^{2\pi} \frac{1}{1+|f(t)|^2}|f'(t)|\mathrm{d}t = \frac{2\pi r}{1+r^2}.$$

Exercise 3.1

Consider the function δ on $\mathbb{D} = \{z \in \mathbb{C} : |z| < 1\}$, defined by setting $\delta(z)$ to be the reciprocal of the Euclidean distance from z to $\mathbb{S}^1 = \partial\mathbb{D}$. Give an explicit formula for $\delta(z)$ in terms of z. For each $0 < r < 1$, let C_r be the Euclidean circle in $\mathbb{D}$ with Euclidean centre 0 and Euclidean radius r, and calculate the length of C_r with respect to the element of arc-length $\delta(z)|\mathrm{d}z|$.

We refer to an element of arc-length of the form $\rho(z)\,|\mathrm{d}z|$ as a *conformal distortion* of the standard element of arc-length $|\mathrm{d}z|$ on $\mathbb{C}$.

Up to this point, we have been considering only differentiable paths. It is both easy and convenient to enlarge the set of paths considered. A path $f : [a, b] \to \mathbb{C}$

is *piecewise differentiable* if f is continuous and if there is a partition of $[a, b]$ into subintervals $[a = a_0, a_1], [a_1, a_2], \ldots, [a_n, a_{n+1} = b]$ so that f is differentiable on each subinterval $[a_k, a_{k+1}]$.

A very natural example of a piecewise differentiable path that is not differentiable comes from considering absolute value. Specifically, consider the path $f : [-1, 1] \to \mathbb{C}$ defined by $f(t) = t + |t|i$. Since $|t|$ is not differentiable at $t = 0$, this is not a differentiable path.

However, on $[-1, 0]$ we have that $|t| = -t$ and hence that $f(t) = t - ti$, which is differentiable. Similarly, on $[0, 1]$ we have that $|t| = t$ and hence that $f(t) = t + ti$, which again is differentiable. So, f is piecewise differentiable on $[-1, 1]$.

Any calculation or operation that we can perform on a differentiable path, we can also perform on a piecewise differentiable path, by expressing it as the concatenation of the appropriate number of differentiable paths. Unless otherwise stated, we assume that all paths are piecewise differentiable.

Exercise 3.2

Calculate the length of the path $f : [-1, 1] \to \mathbb{C}$ given by $f(t) = t + |t|i$ with respect to the element of arc-length $\frac{1}{1+|z|^2}|\mathrm{d}z|$.

One question to consider is what happens to the length of a path $f : [a, b] \to \mathbb{C}$ with respect to the element of arc-length $\rho(z)|\mathrm{d}z|$ when the domain of f is changed. That is, suppose that $h : [\alpha, \beta] \to [a, b]$ is a surjective differentiable function (so that $[a, b] = h([\alpha, \beta])$), and construct a new path by taking the composition $g = f \circ h$. How are $\text{length}_\rho(f)$ and $\text{length}_\rho(g)$ related?

The length of f with respect to $\rho(z)|\mathrm{d}z|$ is the path integral

$$\text{length}_\rho(f) = \int_a^b \rho(f(t))\, |f'(t)|\, \mathrm{d}t,$$

while the length of g with respect to $\rho(z)|\mathrm{d}z|$ is the path integral

$$\begin{aligned} \text{length}_\rho(g) &= \int_\alpha^\beta \rho(g(t))\, |g'(t)|\, \mathrm{d}t \\ &= \int_\alpha^\beta \rho((f \circ h)(t))\, |(f \circ h)'(t)|\, \mathrm{d}t \\ &= \int_\alpha^\beta \rho(f(h(t)))\, |f'(h(t))|\, |h'(t)|\, \mathrm{d}t. \end{aligned}$$

If $h'(t) \geq 0$ for all t in $[\alpha, \beta]$, then $h(\alpha) = a$ and $h(\beta) = b$, and $|h'(t)| = h'(t)$, and so after making the substitution $s = h(t)$, the length of g with respect to

$\rho(z)|\mathrm{d}z|$ becomes

$$\begin{aligned} \text{length}_\rho(g) &= \int_\alpha^\beta \rho(f(h(t)))\,|f'(h(t))|\,|h'(t)|\,\mathrm{d}t \\ &= \int_a^b \rho(f(s))\,|f'(s)|\,\mathrm{d}s = \text{length}_\rho(f). \end{aligned}$$

Similarly, if $h'(t) \leq 0$ for all t in $[\alpha, \beta]$, then $h(\alpha) = b$ and $h(\beta) = a$, and $|h'(t)| = -h'(t)$, and so after making the substitution $s = h(t)$, the length of g with respect to $\rho(z)|\mathrm{d}z|$ becomes

$$\begin{aligned} \text{length}_\rho(g) &= \int_\alpha^\beta \rho(f(h(t)))\,|f'(h(t))|\,|h'(t)|\,\mathrm{d}t \\ &= -\int_b^a \rho(f(s))\,|f'(s)|\,\mathrm{d}s = \text{length}_\rho(f). \end{aligned}$$

So, we have shown that if either $h'(t) \geq 0$ for all t in $[\alpha, \beta]$, or $h'(t) \leq 0$ for all t in $[\alpha, \beta]$, then

$$\text{length}_\rho(f) = \text{length}_\rho(f \circ h),$$

where $f : [a, b] \to \mathbb{C}$ is a piecewise differentiable path and $h : [\alpha, \beta] \to [a, b]$ is differentiable. In this case, we refer to $f \circ h$ as a *reparametrization* of f. Note that reparametrization allows us to choose the domain of definition for a path at will, since we can always find such an h between two intervals.

Though we do not prove it here, the converse of this argument holds as well, namely that $\text{length}_\rho(f) = \text{length}_\rho(f \circ h)$ implies that $h'(t) \geq 0$ for all t or $h'(t) \leq 0$ for all t. This fact is encoded in the following proposition.

Proposition 3.2

Let $f : [a, b] \to \mathbb{C}$ be a piecewise differentiable path, let $[\alpha, \beta]$ be another interval, and let $h : [\alpha, \beta] \to [a, b]$ be a surjective differentiable function. Let $\rho(z)|\mathrm{d}z|$ be an element of arc-length on $\mathbb{C}$. Then

$$\text{length}_\rho(f \circ h) \geq \text{length}_\rho(f)$$

with equality if and only if $h \circ f$ is a reparametrization of f; that is, with equality if and only if either $h'(t) \geq 0$ for all t in $[\alpha, \beta]$, or $h'(t) \leq 0$ for all t in $[\alpha, \beta]$.

Conformal distortions of $|\mathrm{d}z|$ are not the most general form that an element of arc-length on an open subset of $\mathbb{C}$ can take. A more general element of arc-length might have the form $\xi(z, \mathbf{v})|\mathrm{d}z|$, where $\mathbf{v}$ is a vector at z. For a differentiable path $f : [a,b] \to \mathbb{C}$ with non-zero derivative, we interpret this element of arc-length as meaning

$$\int_f \xi(z, \mathbf{v})\, |\mathrm{d}z| = \int_a^b \xi(f(t), f'(t))\, |f'(t)|\, \mathrm{d}t.$$

We will not work with elements of arc-length of this general form, largely because we do not have to. To derive an element of arc-length on $\mathbb{H}$ that is invariant under the action of Möb($\mathbb{H}$), it suffices to work with conformal distortions of $|\mathrm{d}z|$.

3.2 The Element of Arc-length on $\mathbb{H}$

Our goal is to develop a means of measuring hyperbolic length and hyperbolic distance in $\mathbb{H}$. In order to measure hyperbolic length, we need to find an appropriate hyperbolic element of arc-length. Since we wish to measure hyperbolic length, and since we have at hand a group of well-behaved transformations of $\mathbb{H}$, namely Möb($\mathbb{H}$), it seems reasonable to consider those elements of arc-length on $\mathbb{H}$ that are invariant under the action of Möb($\mathbb{H}$).

Let $\rho(z)\, |\mathrm{d}z|$ be an element of arc-length on $\mathbb{H}$ that is a conformal distortion of the standard element of arc-length, so that the length of a piecewise differentiable path $f : [a,b] \to \mathbb{H}$ is given by the integral

$$\text{length}_\rho(f) = \int_f \rho(z)|\mathrm{d}z| = \int_a^b \rho(f(t))\, |f'(t)|\, \mathrm{d}t.$$

While it seems evident that this integral is finite for every path f in $\mathbb{H}$, we show in Proposition 3.7 that this is actually the case.

By the phrase *length is invariant under the action of* Möb($\mathbb{H}$), we mean that for every piecewise differentiable path $f : [a,b] \to \mathbb{H}$ and every element γ of Möb($\mathbb{H}$), we have

$$\text{length}_\rho(f) = \text{length}_\rho(\gamma \circ f).$$

Let us see what conditions this assumption imposes on ρ. We start by taking γ to be an element of Möb$^+$($\mathbb{H}$). Expanding out $\text{length}_\rho(f)$ and $\text{length}_\rho(\gamma \circ f)$, we have

$$\text{length}_\rho(f) = \int_a^b \rho(f(t))\, |f'(t)|\, \mathrm{d}t$$

and

$$\text{length}_\rho(\gamma \circ f) = \int_a^b \rho((\gamma \circ f)(t))\, |(\gamma \circ f)'(t))|\, \mathrm{d}t,$$

and so we are have that

$$\int_a^b \rho(f(t))\, |f'(t)|\, \mathrm{d}t = \int_a^b \rho((\gamma \circ f)(t))\, |(\gamma \circ f)'(t))|\, \mathrm{d}t$$

for every piecewise differentiable path $f : [a,b] \to \mathbb{H}$ and every element γ of $\text{Möb}^+(\mathbb{H})$.

Using the chain rule to expand $(\gamma \circ f)'(t)$ as $(\gamma \circ f)'(t) = \gamma'(f(t))\, f'(t)$, the integral for $\text{length}_\rho(\gamma \circ f)$ becomes

$$\int_a^b \rho(f(t))|f'(t)|\, \mathrm{d}t = \int_a^b \rho((\gamma \circ f)(t))\, |\gamma'(f(t))|\, |f'(t)|\, \mathrm{d}t.$$

Note 3.3

At this point, we need to insert a note about *differentiation of elements of* Möb. Unlike in the case of functions of a single real variable, such as paths, there are two different ways in which to talk about the derivative of an element of Möb.

One is to use complex analysis. That is, we view an element m of Möb as a function from $\overline{\mathbb{C}}$ to $\overline{\mathbb{C}}$, and define its derivative $m'(z)$ (using the usual definition) as

$$m'(z) = \lim_{w \to z} \frac{m(w) - m(z)}{w - z}.$$

Using this definition, all the usual formulae for derivatives hold, such as the product, quotient, and chain rules, and the derivative of an element $m(z) = \frac{az+b}{cz+d}$ of Möb^+ (normalized so that $ad - bc = 1$) is

$$m'(z) = \frac{1}{(cz+d)^2}.$$

This is the definition of differentiable we usually use. These functions are often referred to as *holomorphic* or *analytic*.

However, one disadvantage of this definition is that the derivative of an element of Möb which is not an element of Möb^+ is not defined. In particular, the derivative of $C(z) = \overline{z}$ does not exist.

There is a second way of defining the derivative of an element of Möb, which is to use multivariable calculus. That is, we forget that an element m of Möb is a function of a complex variable and instead view it as a function from $\mathbb{R}^2$ to $\mathbb{R}^2$. In this case, the derivative is no longer a single function, but instead is the 2×2 matrix of partial derivatives. That is, if we write z in terms of its real and imaginary parts as $z = x + iy$ and m in terms of its real and imaginary parts as $m(x,y) = (f(x,y), g(x,y))$, where f and g are real valued functions, then the derivative of m is

$$Dm = \begin{pmatrix} \frac{\partial f}{\partial x} & \frac{\partial f}{\partial y} \\ \frac{\partial g}{\partial x} & \frac{\partial g}{\partial y} \end{pmatrix}.$$

This definition of differentiable is used in the definition of hyperbolic area in Section 5.4.

We distinguish between these two notions of differentiability by referring to the first by saying that *m is differentiable as a function of z* and by referring to the second by saying that *m is differentiable as a function of x and y*. It is true that differentiable as a function of z implies differentiable as a function of x and y, but not conversely. The distinction between these two definitions is one of the topics covered in complex analysis. This concludes Note 3.3.

Getting back to the argument in progress, the condition on $\rho(z)$ then becomes that

$$\int_a^b \rho(f(t))\, |f'(t)|\, \mathrm{d}t = \int_a^b \rho((\gamma \circ f)(t))\, |\gamma'(f(t))|\, |f'(t)|\, \mathrm{d}t$$

for every piecewise differentiable path $f : [a, b] \to \mathbb{H}$ and every element γ of $\text{Möb}^+(\mathbb{H})$. Equivalently, this can be written as

$$\int_a^b (\rho(f(t)) - \rho((\gamma \circ f)(t))\, |\gamma'(f(t))|)\, |f'(t)|\, \mathrm{d}t = 0$$

for every piecewise differentiable path $f : [a, b] \to \mathbb{H}$ and every element γ of $\text{Möb}^+(\mathbb{H})$.

For an element γ of $\text{Möb}^+(\mathbb{H})$, set

$$\mu_\gamma(z) = \rho(z) - \rho(\gamma(z))|\gamma'(z)|,$$

so that the condition on $\rho(z)$ becomes a condition on $\mu_\gamma(z)$, namely that

$$\int_f \mu_\gamma(z)|\mathrm{d}z| = \int_a^b \mu_\gamma(f(t))\, |f'(t)|\, \mathrm{d}t = 0$$

for every piecewise differentiable path $f : [a, b] \to \mathbb{H}$ and every element γ of $\text{Möb}^+(\mathbb{H})$. Note that, since $\rho(z)$ is continuous and γ is differentiable, we have that $\mu_\gamma(z)$ is continuous for every element γ of $\text{Möb}^+(\mathbb{H})$.

This derived condition on $\mu_\gamma(z)$ is more apparently tractable than the original condition on $\rho(z)$, as it is easier to subject to analysis. In particular, making use of this derived condition allows us to remove the requirement that we consider all piecewise differentiable paths in $\mathbb{H}$. This is the content of the following lemma.

Lemma 3.4

Let D be an open subset of $\mathbb{C}$, let $\mu : D \to \mathbb{R}$ be a continuous function, and suppose that $\int_f \mu(z)|dz| = 0$ for every piecewise differentiable path $f : [a, b] \to D$. Then, $\mu \equiv 0$.

The proof of Lemma 3.4 is by contradiction, so suppose there exists a point $z \in D$ at which $\mu(z) \neq 0$. Replacing μ by $-\mu$ if necessary, we may assume that $\mu(z) > 0$.

The hypothesis that μ is continuous yields that for each $\varepsilon > 0$, there exists $\delta > 0$ so that $U_\delta(z) \subset D$ and $w \in U_\delta(z)$ implies that $\mu(w) \in U_\varepsilon(\mu(z))$, where

$$U_\delta(z) = \{u \in \mathbb{C} : |u - z| < \delta\} \text{ and } U_\varepsilon(t) = \{s \in \mathbb{R} : |s - t| < \varepsilon\}.$$

Taking $\varepsilon = \frac{1}{3}|\mu(z)|$, we see that there exists $\delta > 0$ so that $w \in U_\delta(z)$ implies that $\mu(w) \in U_\varepsilon(\mu(z))$. Using the triangle inequality and the fact that $\mu(z) > 0$, this implies that $\mu(w) > 0$ for all $w \in U_\delta(z)$.

We now choose a specific non-constant piecewise differentiable path, namely the path $f : [0, 1] \to U_\delta(z)$ given by

$$f(t) = z + \frac{1}{3}\delta t.$$

Observe that $\mu(f(t)) > 0$ for all t in $[0, 1]$, since $f(t) \in U_\delta(z)$ for all t in $[0, 1]$. In particular, we have that $\int_f \mu(z)|dz| > 0$, which gives the desired contradiction. This completes the proof of Lemma 3.4.

Recall that we are assuming that length is invariant under the action of $\text{Möb}^+(\mathbb{H})$, which implies that $\int_f \mu_\gamma(z)|dz| = 0$ for every piecewise differentiable path $f : [a, b] \to \mathbb{H}$ and every element γ of $\text{Möb}^+(\mathbb{H})$. Applying Lemma 3.4 to $\mu_\gamma(z)$, this leads us to the conclusion that

$$\mu_\gamma(z) = \rho(z) - \rho(\gamma(z))|\gamma'(z)| = 0$$

for every $z \in \mathbb{H}$ and every element γ of $\text{Möb}^+(\mathbb{H})$.

In order to simplify our analysis, we consider how μ_γ behaves under composition of elements of $\text{Möb}^+(\mathbb{H})$. Let γ and φ be two elements in $\text{Möb}^+(\mathbb{H})$. Calculating, we see that

$$\begin{aligned}
\mu_{\gamma\circ\varphi}(z) &= \rho(z) - \rho((\gamma \circ \varphi)(z))|(\gamma \circ \varphi)'(z)| \\
&= \rho(z) - \rho((\gamma \circ \varphi)(z))|\gamma'(\varphi(z))||\varphi'(z)| \\
&= \rho(z) - \rho(\varphi(z))|\varphi'(z)| + \rho(\varphi(z))|\varphi'(z)| \\
&\quad - \rho((\gamma \circ \varphi)(z))|\gamma'(\varphi(z))||\varphi'(z)| \\
&= \mu_\varphi(z) + \mu_\gamma(\varphi(z))|\varphi'(z)|.
\end{aligned}$$

In particular, if $\mu_\gamma \equiv 0$ for every γ in a generating set for $\text{Möb}^+(\mathbb{H})$, then $\mu_\gamma \equiv 0$ for every element γ of $\text{Möb}^+(\mathbb{H})$. We saw in Exercise 2.37 that there exists a generating set for $\text{Möb}^+(\mathbb{H})$ consisting of the transformations $m(z) = az + b$ for a, $b \in \mathbb{R}$ and $a > 0$, together with the transformation $K(z) = -\frac{1}{z}$.

Again, we are putting off consideration of $B(z) = -\overline{z}$ until later.

So, it suffices to analyze our condition on μ_γ, and hence on ρ, for the elements of this generating set. We consider these generators one at a time.

We first consider the generator $\gamma(z) = z + b$ for $b \in \mathbb{R}$. Since $\gamma'(z) = 1$ for every $z \in \mathbb{H}$, the condition imposed on $\rho(z)$ is that

$$0 \equiv \mu_\gamma(z) = \rho(z) - \rho(\gamma(z))|\gamma'(z)| = \rho(z) - \rho(z+b)$$

for every $z \in \mathbb{H}$ and every $b \in \mathbb{R}$. That is,

$$\rho(z) = \rho(z+b)$$

for every $z \in \mathbb{H}$ and every $b \in \mathbb{R}$. In particular, $\rho(z)$ depends only on the imaginary part $y = \text{Im}(z)$ of $z = x + iy$.

To see this explicitly, suppose that $z_1 = x_1 + iy$ and $z_2 = x_2 + iy$ have the same imaginary part, and write $z_2 = z_1 + (x_2 - x_1)$. Since $x_2 - x_1$ is real, we have that $\rho(z_2) = \rho(z_1)$.

Hence, we may view ρ as a real-valued function of the single real variable $y = \text{Im}(z)$. Explicitly, consider the real-valued function $r : (0, \infty) \to (0, \infty)$ given by $r(y) = \rho(iy)$, and note that $\rho(z) = r(\text{Im}(z))$ for every $z \in \mathbb{H}$.

We now consider the generator $\gamma(z) = az$ for $a > 0$. Since $\gamma'(z) = a$ for every $z \in \mathbb{H}$, the condition imposed on $\rho(z)$ is that

$$0 \equiv \mu_\gamma(z) = \rho(z) - \rho(\gamma(z))|\gamma'(z)| = \rho(z) - a\rho(az)$$

for every $z \in \mathbb{H}$ and every $a > 0$. That is,

$$\rho(z) = a\rho(az)$$

for every $z \in \mathbb{H}$ and every $a > 0$. In particular, we have that

$$r(y) = ar(ay)$$

for every $y > 0$ and every $a > 0$. Interchanging the roles of a and y, we see that $r(a) = yr(ay)$. Dividing through by y, we obtain

$$r(ay) = \frac{1}{y}\, r(a).$$

Taking $a = 1$, this yields that

$$r(y) = \frac{1}{y}\, r(1),$$

and so r is completely determined by its value at 1.

Recalling the definition of r, we have that the invariance of length under $\text{Möb}^+(\mathbb{H})$ implies that $\rho(z)$ has the form

$$\rho(z) = r(\text{Im}(z)) = \frac{c}{\text{Im}(z)},$$

where c is an arbitrary positive constant.

Exercise 3.3

For a real number $\lambda > 0$, let A_λ be the Euclidean line segment joining $-1 + i\lambda$ to $1 + i\lambda$, and let B_λ be the hyperbolic line segment joining $-1 + i\lambda$ to $1 + i\lambda$. Calculate the lengths of A_λ and B_λ with respect to the element of arc-length $\frac{c}{\text{Im}(z)}\,|\text{d}z|$.

Note that the derivation $\rho(z)$ we have just performed does not use all the generators of $\text{Möb}(\mathbb{H})$. One question to be addressed is whether this form for $\rho(z)$ is consistent with lengths of paths being assumed to be invariant under both $K(z) = -\frac{1}{z}$ and $B(z) = -\overline{z}$.

Exercise 3.4

Check that the length of a piecewise differentiable path $f : [a, b] \to \mathbb{H}$ calculated with respect to the element of arc-length $\frac{c}{\text{Im}(z)}\,|\text{d}z|$ is invariant under both $K(z) = -\frac{1}{z}$ and $B(z) = -\overline{z}$. (Note that for $B(z)$, we cannot use the argument just given, as $B'(z)$ is not defined; instead, proceed directly by first evaluating the composition $B \circ f$ and then differentiating it as a path.)

Assuming the result of Exercise 3.4, we have proven the following theorem.

Theorem 3.5

For every positive constant c, the element of arc-length

$$\frac{c}{\text{Im}(z)}\,|\text{d}z|$$

on $\mathbb{H}$ is invariant under the action of $\text{Möb}(\mathbb{H})$.

That is, for every piecewise differentiable path $f : [a, b] \to \mathbb{H}$ and every element γ of $\text{Möb}(\mathbb{H})$, we have that

$$\text{length}_\rho(f) = \text{length}_\rho(\gamma \circ f).$$

However, nothing we have done to this point has given us a way of determining a specific value of c. In fact, it is not possible to specify the value of c using solely the action of $\text{Möb}(\mathbb{H})$. To avoid carrying c through all our calculations, we set $c = 1$.

Definition 3.6

For a piecewise differentiable path $f : [a, b] \to \mathbb{H}$, we define the *hyperbolic length* of f to be

$$\text{length}_{\mathbb{H}}(f) = \int_f \frac{1}{\text{Im}(z)}\, |\text{d}z| = \int_a^b \frac{1}{\text{Im}(f(t))}\, |f'(t)|\, \text{d}t.$$

There are some paths whose hyperbolic length is straightforward to calculate. As an example, take $0 < a < b$ and consider the path $f : [a, b] \to \mathbb{H}$ given by $f(t) = it$. The image $f([a, b])$ of $[a, b]$ under f is the segment of the positive imaginary axis between ai and bi. Since $\text{Im}(f(t)) = t$ and $|f'(t)| = 1$, we see that

$$\text{length}_{\mathbb{H}}(f) = \int_f \frac{1}{\text{Im}(z)}|\text{d}z| = \int_a^b \frac{1}{t}\text{d}t = \ln\left[\frac{b}{a}\right].$$

There are also paths whose hyperbolic length is more difficult to calculate.

Exercise 3.5

For each natural number n, write down the integral for the hyperbolic length of the path $f_n : [0, 1] \to \mathbb{H}$ given by

$$f_n(t) = t + i(t^n + 1).$$

Exercise 3.6

For each of the paths f_n defined in Exercise 3.5, make a conjecture about the behaviour of the hyperbolic length of $\gamma_n = f_n([0, 1])$ as $n \to \infty$, and calculate the putative limit of the hyperbolic length of γ_n as $n \to \infty$.

There is one subtlety that we need to mention regarding hyperbolic length before going on, namely that piecewise differentiable paths in $\mathbb{H}$ have finite hyperbolic length.

Proposition 3.7

Let $f : [a,b] \to \mathbb{H}$ be a piecewise differentiable path. Then, the hyperbolic length $\text{length}_{\mathbb{H}}(f)$ of f is finite.

The proof of Proposition 3.7 is an immediate consequence of the fact that there exists a constant $B > 0$ so that the image $f([a,b])$ of $[a,b]$ under f is contained in the subset

$$K_B = \{z \in \mathbb{H} \mid \text{Im}(z) \geq B\}$$

of $\mathbb{H}$. This fact follows from that fact that $[a,b]$, and hence $f([a,b])$, are *compact*, a concept discussed in more detail in Section 3.7.

Given that $f([a,b])$ is contained in K_B, we can estimate the integral giving the hyperbolic length of f. We first note that by the definition of piecewise differentiable, there is a partition P of $[a,b]$ into subintervals

$$P = \{[a = a_0, a_1], [a_1, a_2], \ldots, [a_n, a_{n+1} = b]\}$$

so that f is differentiable on each subinterval $[a_k, a_{k+1}]$.

In particular, its derivative f' is continuous on each subinterval. By the extreme value theorem for a continuous function on a closed interval, there then exists for each k a number A_k so that

$$|f'(t)| \leq A_k \text{ for all } t \in [a_k, a_{k+1}].$$

Let A be the maximum of $A_0, \ldots, A_n$. Then, we have that

$$\text{length}_{\mathbb{H}}(f) = \int_a^b \frac{1}{\text{Im}(f(t))} \, |f'(t)| \, \mathrm{d}t \leq \int_a^b \frac{1}{B} \, A \, \mathrm{d}t = \frac{A}{B} \, (b - a),$$

which is finite. This completes the proof of Proposition 3.7.

We close this section by noting that the proof of Proposition 3.7 gives a crude way of estimating an upper bound for the hyperbolic length of a path in $\mathbb{H}$.

3.3 Path Metric Spaces

We now know how to calculate the hyperbolic length of every piecewise differentiable path in $\mathbb{H}$, namely by integrating the hyperbolic element of arc-length $\frac{1}{\text{Im}(z)} \, |\mathrm{d}z|$ along the path. We are now able to apply a general construction to pass from calculating hyperbolic lengths of paths in $\mathbb{H}$ to getting a hyperbolic metric on $\mathbb{H}$.

We begin by recalling the definition of a *metric*. Roughly, a metric on a set X is a means of assigning a distance between any pair of points of X. We give only a very brief and non-comprehensive description of metrics in this section. For a more detailed discussion of metrics, the interested reader should consult a textbook on point-set topology, such as the book of Munkres [19].

Definition 3.8

A *metric* on a set X is a function

$$\mathrm{d} : X \times X \to \mathbb{R}$$

satisfying three conditions:

1 $\mathrm{d}(x, y) \geq 0$ for all x, $y \in X$, and $\mathrm{d}(x, y) = 0$ if and only if $x = y$;

2 $\mathrm{d}(x, y) = \mathrm{d}(y, x)$; and

3 $\mathrm{d}(x, z) \leq \mathrm{d}(x, y) + \mathrm{d}(y, z)$ (the triangle inequality).

If d is a metric on X, we often refer to the *metric space* (X, d). The notion of a metric is very general, but it is good to keep in mind that we have already encountered several examples of metrics.

One example is the standard metric on $\mathbb{R}$ and $\mathbb{C}$ given by absolute value. On $\mathbb{C}$, this metric is given explicitly by the function

$$\mathrm{n} : \mathbb{C} \times \mathbb{C} \to \mathbb{R}, \text{ where } \mathrm{n}(z, w) = |z - w|.$$

The three conditions defining a metric on a general set can be thought of as an abstraction of the familiar properties of this function n.

A more complicated example is the metric on the Riemann sphere $\overline{\mathbb{C}}$ given by the function

$$\mathrm{s} : \overline{\mathbb{C}} \times \overline{\mathbb{C}} \to \mathbb{R},$$

where

$$\mathrm{s}(z, w) = \frac{2|z - w|}{\sqrt{(1 + |z|^2)(1 + |w|^2)}}$$

for z, $w \in \mathbb{C}$, and

$$\mathrm{s}(z, \infty) = \mathrm{s}(\infty, z) = \frac{2}{\sqrt{1 + |z|^2}}$$

for $z \in \mathbb{C}$.

The proof that s is a metric on $\overline{\mathbb{C}}$ makes use of stereographic projection. These formulae are the expressions, in terms of the coordinate on $\overline{\mathbb{C}}$, of the Eulidean distances in $\mathbb{R}^3$ between the corresponding points on $\mathbb{S}^2$.

Note 3.9

Note that whenever we have a metric d on a space X, we can mimic in X the definitions of open and closed sets that we have in $\mathbb{C}$ and in $\overline{\mathbb{C}}$, and so we have notions of convergence of sequences in (X, d), and continuity of functions whose domain or range is the metric space (X, d).

Specifically, in the metric space (X, d), we can define the open disc $U_\varepsilon(x)$ of radius $\varepsilon > 0$ centred at a point x as

$$U_\varepsilon(x) = \{y \in X \, : \, \mathrm{d}(x,y) < \varepsilon\}.$$

Then, a subset A of X is *open* if for every $x \in A$, there exists some $\varepsilon > 0$ so that $U_\varepsilon(x) \subset A$; a subset B of X is *closed* if its complement $X - B$ is open.

A sequence $\{x_n\}$ of points of X *converges* to a point x of X if for every $\varepsilon > 0$, there exists some $N > 0$ so that $x_n \in U_\varepsilon(x)$ for all $n > N$.

We can also define *continuity* of functions between metric spaces. If (X, d_X) and (Y, d_Y) are two metric spaces and if $f : X \to Y$ is a function, then f is *continuous at a point x of X* if given $\varepsilon > 0$, there exists $\delta > 0$ so that $f(U_\delta(x)) \subset U_\varepsilon(f(x))$. We say that f is *continuous* if it is continuous at every point of X.

One example of a continuous function comes from the metric itself. Fix a point $z \in X$, and consider the function $f : X \to \mathbb{R}$ given by $f(x) = \mathrm{d}(z, x)$. Then, this function f is continuous. We actually make use of the continuity of this function in Section 5.1.

Hence, we can carry over a great many of the concepts familiar from our knowledge of $\mathbb{C}$ and $\overline{\mathbb{C}}$ to any metric space. This concludes Note 3.9.

There is one more example of a metric space which will be very important to us in our study of the hyperbolic plane. Let X be a set in which we know how to measure lengths of paths. Specifically, for each pair x and y of points in X, let $\Gamma[x, y]$ be a non-empty collection of paths $f : [a, b] \to X$ satisfying $f(a) = x$ and $f(b) = y$, and assume that to each path f in $\Gamma[x, y]$ we can associate a non-negative real number length(f), which we refer to as the *length of f*.

As an example to keep in mind, take X to be the upper half-plane $\mathbb{H}$, and take $\Gamma[x, y]$ to be the set of all piecewise differentiable paths $f : [a, b] \to \mathbb{H}$ with $f(a) = x$ and $f(b) = y$, where the length of each path f in $\Gamma[x, y]$ is just the hyperbolic length $\mathrm{length}_{\mathbb{H}}(f)$ of f.

Consider the function $\mathrm{d} : X \times X \to \mathbb{R}$ defined by taking the infimum

$$\mathrm{d}(x, y) = \inf\{\mathrm{length}(f) \, : \, f \in \Gamma[x, y]\}.$$

There are several questions to ask about the construction of this function d. One question is what conditions on the definition of length are needed to determine whether d defines a metric on X. In order to avoid technical difficulties, we do not consider this question in general, as we are most interested in the case of the metric on $\mathbb{H}$ coming from hyperbolic lengths of paths, which we consider in detail in Section 3.4.

A second question, assuming that d does indeed define a metric on X, is whether there necessarily exist *distance realizing paths* in X. That is, given a pair x and y of points in X, does there necessarily exist a path f in $\Gamma[x, y]$ for which $\text{length}(f) = \text{d}(x, y)$.

As mentioned above, we consider both of these questions in detail for the upper half-plane $\mathbb{H}$ in Section 3.4. As an illustrative case, though, we consider some general properties of this construction for the case $X = \mathbb{C}$. We do not give any specific details, since they are very similar to the details given for $\mathbb{H}$.

For each pair x and y of points of $\mathbb{C}$, let $\Gamma[x, y]$ be the set of all piecewise differentiable paths $f : [a, b] \to \mathbb{C}$ with $f(a) = x$ and $f(b) = y$, and let $\text{length}(f)$ be the usual Euclidean length of f. In this case, since the shortest Euclidean distance between two points is along a Euclidean line, which can be parametrized by a differentiable path, we see that

$$\text{d}(x, y) = \inf\{\text{length}(f) \, : \, f \in \Gamma[x, y]\} = \text{n}(x, y).$$

Note that in this case, this construction of a function on $\mathbb{C} \times \mathbb{C}$ by taking the infimum of the lengths of paths gives rise to the standard metric $\text{n}(\cdot, \cdot)$ on $\mathbb{C}$. In particular there is always a path in $\Gamma[x, y]$ realizing the Euclidean distance $\text{d}(x, y)$ between x and y, namely the Euclidean line segment joining x to y.

There is a related example that illustrates one of the difficulties that can arise. Let $X = \mathbb{C} - \{0\}$ be the punctured plane, and for each pair of points x and y of X, let $\Gamma[x, y]$ be the set of all piecewise differentiable paths $f : [a, b] \to X$ with $f(a) = x$ and $f(b) = y$.

In this case, we can bring what we know about the behaviour of $(\mathbb{C}, \text{n})$ to bear in our analysis of X. Again, this construction of a function on $X \times X$ by taking the infimum of the lengths of paths gives rise to the metric $\text{n}(x, y) = |x - y|$ on X.

However, we no longer have that there always exists a path in $\Gamma[x, y]$ realizing the Euclidean distance between x and y. Specifically, consider the two points 1 and -1. The Euclidean line segment joining 1 to -1 passes through 0, and so is not a path in X. Every other path joining 1 to -1 has length strictly greater than $\text{n}(1, -1) = 2$.

So, recall that we are working in a set X in which we know how to measure lengths of paths. For each pair x and y of points of X, there exists a non-empty collection $\Gamma[x,y]$ of paths $f : [a,b] \to X$ satisfying $f(a) = x$ and $f(b) = y$, and for each path f in $\Gamma[x,y]$ we denote the length of f by length(f).

Suppose that in addition X is a metric space with metric d. We say that (X, d) is a *path metric space* if for each pair of points x and y of X we have that

$$\mathrm{d}(x,y) = \inf\{\mathrm{length}(f) \ : \ f \in \Gamma[x,y]\},$$

and for each pair of points x and y of X we have that there exists a distance realizing path in $\Gamma[x,y]$, which is a path f in $\Gamma[x,y]$ satisfying

$$\mathrm{d}(x,y) = \mathrm{length}(f).$$

We note that this definition of path metric space is stronger than the standard definition, as we require the existence of a distance realizing path.

Of the metric spaces mentioned in this section, we have with this definition that $(\mathbb{C}, \mathrm{n})$ and $(\overline{\mathbb{C}}, \mathrm{s})$ are path metric spaces, while $(\mathbb{C} - \{0\}, \mathrm{n})$ is not.

3.4 From Arc-length to Metric

We are now ready to prove that $\mathbb{H}$ is a path metric space. The proof of this fact takes up the bulk of this section.

For each pair of points x and y of $\mathbb{H}$, let $\Gamma[x,y]$ denote the set of all piecewise differentiable paths $f : [a,b] \to \mathbb{H}$ with $f(a) = x$ and $f(b) = y$.

Since we can parametrize the hyperbolic line segment joining x to y by a piecewise differentiable path, we see that $\Gamma[x,y]$ is non-empty. Also, by Proposition 3.7, we know that every path f in $\Gamma[x,y]$ has finite hyperbolic length $\mathrm{length}_{\mathbb{H}}(f)$.

Consider the function

$$\mathrm{d}_{\mathbb{H}} : \mathbb{H} \times \mathbb{H} \to \mathbb{R}$$

defined by

$$\mathrm{d}_{\mathbb{H}}(x,y) = \inf\{\mathrm{length}_{\mathbb{H}}(f) \ : \ f \in \Gamma[x,y]\}.$$

In anticipation of the proof of Theorem 3.10, we refer to $\mathrm{d}_{\mathbb{H}}(x,y)$ as the *hyperbolic distance* between x and y.

Theorem 3.10

$(\mathbb{H}, \mathrm{d}_{\mathbb{H}})$ is a path metric space. Moreover, the distance realizing path in $\Gamma[x,y]$ is a parametrization of the hyperbolic line segment joining x to y.

Since the hyperbolic length of a path is invariant under the action of $\text{Möb}(\mathbb{H})$, we have the following useful observation.

Proposition 3.11

For every element γ of $\text{Möb}(\mathbb{H})$ and for every pair x and y of points of $\mathbb{H}$, we have that

$$\mathrm{d}_{\mathbb{H}}(x, y) = \mathrm{d}_{\mathbb{H}}(\gamma(x), \gamma(y)).$$

We begin by observing that $\{\gamma \circ f \,:\, f \in \Gamma[x,y]\} \subset \Gamma[\gamma(x), \gamma(y)]$. To see this, take a path $f : [a,b] \to \mathbb{H}$ in $\Gamma[x,y]$, so that $f(a) = x$ and $f(b) = y$. Since $\gamma \circ f(a) = \gamma(x)$ and $\gamma \circ f(b) = \gamma(y)$, we have that $\gamma \circ f$ lies in $\Gamma[\gamma(x), \gamma(y)]$.

Since $\text{length}_{\mathbb{H}}(f)$ is invariant under the action of $\text{Möb}(\mathbb{H})$, we have that

$$\text{length}_{\mathbb{H}}(\gamma \circ f) = \text{length}_{\mathbb{H}}(f)$$

for every path f in $\Gamma[x,y]$, and so

$$\begin{aligned}
\mathrm{d}_{\mathbb{H}}(\gamma(x), \gamma(y)) &= \inf\{\text{length}_{\mathbb{H}}(g) \,:\, g \in \Gamma[\gamma(x), \gamma(y)]\} \\
&\le \inf\{\text{length}_{\mathbb{H}}(\gamma \circ f) \,:\, f \in \Gamma[x,y]\} \\
&\le \inf\{\text{length}_{\mathbb{H}}(f) \,:\, f \in \Gamma[x,y]\} = \mathrm{d}_{\mathbb{H}}(x,y).
\end{aligned}$$

Since γ is invertible and γ^{-1} is an element of $\text{Möb}(\mathbb{H})$, we may repeat the argument just given to see that

$$\{\gamma^{-1} \circ g \mid g \in \Gamma[\gamma(x), \gamma(y)]\} \subset \Gamma[x,y],$$

and hence that

$$\begin{aligned}
\mathrm{d}_{\mathbb{H}}(x, y) &= \inf\{\text{length}_{\mathbb{H}}(f) \,:\, f \in \Gamma[x,y]\} \\
&\le \inf\{\text{length}_{\mathbb{H}}(\gamma^{-1} \circ g) \,:\, g \in \Gamma[\gamma(x), \gamma(y)]\} \\
&\le \inf\{\text{length}_{\mathbb{H}}(g) \,:\, g \in \Gamma[\gamma(x), \gamma(y)]\} = \mathrm{d}_{\mathbb{H}}(\gamma(x), \gamma(y)).
\end{aligned}$$

In particular, this yields that $\mathrm{d}_{\mathbb{H}}(x,y) = \mathrm{d}_{\mathbb{H}}(\gamma(x), \gamma(y))$. This completes the proof of Proposition 3.11.

In order to show that $\mathrm{d}_{\mathbb{H}}$ does indeed define a metric, we need to show that $\mathrm{d}_{\mathbb{H}}$ satisfies the three conditions given at the beginning of Section 3.3.

Let $f : [a,b] \to \mathbb{H}$ be a path in $\Gamma[x,y]$, and recall the definition of $\text{length}_{\mathbb{H}}(f)$:

$$\text{length}_{\mathbb{H}}(f) = \int_f \frac{1}{\text{Im}(z)}\,|\mathrm{d}z| = \int_a^b \frac{1}{\text{Im}(f(t))}\,|f'(t)|\mathrm{d}t.$$

Since the integrand is always non-negative, it is immediate that the integral is non-negative. Since $\text{length}_{\mathbb{H}}(f)$ is non-negative for every path f in $\Gamma[x,y]$, the infimum $\mathrm{d}_{\mathbb{H}}(x,y)$ of these integrals is non-negative.

This shows that the first part of Condition 1 of the definition of a metric is satisfied by $\mathrm{d}_{\mathbb{H}}$. For reasons that will become clear at the time, the proof that $\mathrm{d}_{\mathbb{H}}$ satisfies the second part of Condition 1 is postponed to later in the section.

We now consider Condition 2 of the definition of a metric. We need to compare the lengths of paths in $\Gamma[x,y]$ and $\Gamma[y,x]$. Let $f : [a,b] \to \mathbb{H}$ be a path in $\Gamma[x,y]$, and consider the composition of f with the function $h : [b,a] \to [a,b]$ given by $h(t) = a + b - t$. Note that $h'(t) = -1$.

It is evident that $f \circ h$ lies in $\Gamma[y,x]$, since $(f \circ h)(a) = f(b) = y$ and $(f \circ h)(b) = f(a) = x$. Moreover, direct calculation yields that

$$\begin{aligned}
\text{length}_{\mathbb{H}}(f \circ h) &= \int_{f\circ h} \frac{1}{\text{Im}(z)}\, |\mathrm{d}z| \\
&= \int_b^a \frac{1}{\text{Im}((f \circ h)(t))}\, |(f \circ h)'(t)|\, \mathrm{d}t \\
&= \int_b^a \frac{1}{\text{Im}(f(h(t)))}\, |f'(h(t))|\, |h'(t)|\, \mathrm{d}t \\
&= -\int_b^a \frac{1}{\text{Im}(f(s))}\, |f'(s)|\, \mathrm{d}s \\
&= \int_a^b \frac{1}{\text{Im}(f(s))}\, |f'(s)|\, \mathrm{d}s = \text{length}_{\mathbb{H}}(f).
\end{aligned}$$

So, every path in $\Gamma[x,y]$ gives rise to a path in $\Gamma[y,x]$ of equal length, by composing with the appropriate h. Using the same argument, every path in $\Gamma[y,x]$ gives rise to a path in $\Gamma[x,y]$ of equal length.

In particular, we see that the two sets of hyperbolic lengths

$$\{\text{length}_{\mathbb{H}}(f) \;:\; f \in \Gamma[x,y]\} \text{ and } \{\text{length}_{\mathbb{H}}(g) \;:\; g \in \Gamma[y,x]\}$$

are equal. Hence, they have the same infimum, and so $\mathrm{d}_{\mathbb{H}}(x,y) = \mathrm{d}_{\mathbb{H}}(y,x)$. This completes the proof that Condition 2 of the definition of a metric is satisfied by $\mathrm{d}_{\mathbb{H}}$.

We now consider Condition 3 of the definition of a metric, the triangle inequality. To this end, let x, y, and z be points in $\mathbb{H}$.

Conceptually, the simplest proof would be for us to choose a path $f : [a,b] \to \mathbb{H}$ in $\Gamma[x,y]$ with $\text{length}_{\mathbb{H}}(f) = \mathrm{d}_{\mathbb{H}}(x,y)$ and a path $g : [b,c] \to \mathbb{H}$ in $\Gamma[y,z]$ with $\text{length}_{\mathbb{H}}(g) = \mathrm{d}_{\mathbb{H}}(y,z)$. The concatenation $h : [a,c] \to \mathbb{H}$ of f and g would then lie in $\Gamma[x,z]$. Moreover, we would have the desired inequality

$$\mathrm{d}_{\mathbb{H}}(x,z) \leq \text{length}_{\mathbb{H}}(h) = \text{length}_{\mathbb{H}}(f) + \text{length}_{\mathbb{H}}(g) = \mathrm{d}_{\mathbb{H}}(x,y) + \mathrm{d}_{\mathbb{H}}(y,z).$$

We note here that the concatenation of piecewise differentiable paths is again piecewise differentiable, while the concatenation of differentiable paths is not necessarily differentiable. This is one reason to consider piecewise differentiable paths instead of differentiable paths.

Unfortunately, we do not yet know that there always exists a path realizing the hyperbolic distance between a pair of points. We consider this question later in the section. For now, we take a route that is slightly more roundabout. We use proof by contradiction.

Suppose that Condition 3, the triangle inequality, does not hold for $\mathrm{d}_{\mathbb{H}}$. That is, suppose that there exist distinct points x, y, and z in $\mathbb{H}$ so that

$$\mathrm{d}_{\mathbb{H}}(x,z) > \mathrm{d}_{\mathbb{H}}(x,y) + \mathrm{d}_{\mathbb{H}}(y,z).$$

Set

$$\varepsilon = \mathrm{d}_{\mathbb{H}}(x,z) - (\mathrm{d}_{\mathbb{H}}(x,y) + \mathrm{d}_{\mathbb{H}}(y,z)).$$

Since $\mathrm{d}_{\mathbb{H}}(x,y) = \inf\{\mathrm{length}_{\mathbb{H}}(f) : f \in \Gamma[x,y]\}$, there exists a path $f : [a,b] \to \mathbb{H}$ in $\Gamma[x,y]$ with

$$\mathrm{length}_{\mathbb{H}}(f) - \mathrm{d}_{\mathbb{H}}(x,y) < \frac{1}{2}\varepsilon.$$

Similarly, there exists a path $g : [b,c] \to \mathbb{H}$ in $\Gamma[y,z]$ with

$$\mathrm{length}_{\mathbb{H}}(g) - \mathrm{d}_{\mathbb{H}}(y,z) < \frac{1}{2}\varepsilon.$$

Recall that we are able to choose the domains of definition of f and g at will, using our discussion of reparametrization in Section 3.1.

Let $h : [a,c] \to \mathbb{H}$ be the concatenation of f and g.

Since the concatenation of two piecewise differentiable paths is again piecewise differentiable, we have that h lies in $\Gamma[x,z]$. Calculating, we see that

$$\mathrm{length}_{\mathbb{H}}(h) = \mathrm{length}_{\mathbb{H}}(f) + \mathrm{length}_{\mathbb{H}}(g) < \mathrm{d}_{\mathbb{H}}(x,y) + \mathrm{d}_{\mathbb{H}}(y,z) + \varepsilon.$$

Since $\mathrm{d}_{\mathbb{H}}(x,z) \leq \mathrm{length}_{\mathbb{H}}(h)$ by definition of $\mathrm{d}_{\mathbb{H}}$, this gives that

$$\mathrm{d}_{\mathbb{H}}(x,z) < \mathrm{d}_{\mathbb{H}}(x,y) + \mathrm{d}_{\mathbb{H}}(y,z) + \varepsilon,$$

which contradicts the construction of ε. This completes the proof that Condition 3 of the definition of a metric is satisfied by $\mathrm{d}_{\mathbb{H}}$.

There are two things that remain to be checked before we can conclude that $(\mathbb{H}, \mathrm{d}_{\mathbb{H}})$ is a path metric space. We need to show that $\mathrm{d}_{\mathbb{H}}$ satisfies the second part of Condition 1 of the definition of a metric, and we need to show that there exists a path realizing the hyperbolic distance $d_{\mathbb{H}}(x, y)$ between any pair of points x and y of $\mathbb{H}$.

The approach we take comes from the observation that if there exists a path in $\mathbb{H}$ realizing the hyperbolic distance between any pair of points of $\mathbb{H}$, then this implies that $d_{\mathbb{H}}(x, y) > 0$ for $x \neq y$, since the lengths of non-constant paths are positive. Thus, we have the second part of Condition 1 for free.

So, let x and y be a pair of distinct points of $\mathbb{H}$, and let ℓ be the hyperbolic line passing through x and y. We begin by simplifying the situation. From our work in Section 2.9, specifically Exercise 2.39, we know that there exists an element γ of Möb($\mathbb{H}$) so that $\gamma(\ell)$ is the positive imaginary axis in $\mathbb{H}$.

Write $\gamma(x) = \mu i$ and $\gamma(y) = \lambda i$. If $\lambda < \mu$, then use $K \circ \gamma$ instead of γ, where $K(z) = -\frac{1}{z}$, so that $\mu < \lambda$.

Since hyperbolic lengths of paths in $\mathbb{H}$ calculated with respect to the hyperbolic element of arc-length $\frac{1}{\text{Im}(z)}\,|dz|$ are invariant under the action of Möb($\mathbb{H}$), we have that $d_{\mathbb{H}}(x, y) = d_{\mathbb{H}}(\gamma(x), \gamma(y))$. So, we have reduced ourselves to showing that there exists a distance realizing path between μi and λi for $\mu < \lambda$.

We begin this calculation by calculating the hyperbolic length of a specific path, namely the path $f_0 : [\mu, \lambda] \to \mathbb{H}$ defined by $f_0(t) = ti$. The image of f_0 is the hyperbolic line segment joining μi and λi. Since we expect the shortest hyperbolic distance between two points to be along a hyperbolic line, this path seems to be a reasonable choice to be the shortest path in $\Gamma[\mu i, \lambda i]$.

To calculate the length of f_0, we observe that $\text{Im}(f_0(t)) = t$ and $|f_0'(t)| = 1$, and so

$$\text{length}_{\mathbb{H}}(f_0) = \int_{\mu}^{\lambda} \frac{1}{t}\, dt = \ln\left[\frac{\lambda}{\mu}\right].$$

Now, let $f : [a, b] \to \mathbb{H}$ be any path in $\Gamma[\mu i, \lambda i]$. We complete the proof that $\text{length}_{\mathbb{H}}(f_0) = d_{\mathbb{H}}(\mu i, \lambda i)$ by showing that $\text{length}_{\mathbb{H}}(f_0) \leq \text{length}_{\mathbb{H}}(f)$. We do this in several stages, at each stage modifying f to decrease its hyperbolic length, and arguing that it becomes no shorter than f_0 through these modifications.

Write $f(t) = x(t) + y(t)i$. The first modification of f is to ignore the real part. That is, consider the path $g : [a, b] \to \mathbb{H}$ defined by setting

$$g(t) = \text{Im}(f(t))i = y(t)i.$$

Since $g(a) = f(a) = \mu i$ and $g(b) = f(b) = \lambda i$, we see that g lies in $\Gamma[\mu i, \lambda i]$.

Using that $(x'(t))^2 \geq 0$ for all t and that $\text{Im}(g(t)) = \text{Im}(f(t)) = y(t)$, we have that

$$\text{length}_{\mathbb{H}}(g) = \int_a^b \frac{1}{\text{Im}(g(t))}\, |g'(t)|\, dt$$

$$\begin{aligned} &= \int_a^b \frac{1}{y(t)} \sqrt{(y'(t))^2}\,\mathrm{d}t \\ &\leq \int_a^b \frac{1}{y(t)} \sqrt{(x'(t))^2 + (y'(t))^2}\,\mathrm{d}t \\ &\leq \int_a^b \frac{1}{\mathrm{Im}(f(t))}\,|f'(t)|\,\mathrm{d}t = \mathrm{length}_{\mathbb{H}}(f). \end{aligned}$$

So, given any path f in $\Gamma[\mu i, \lambda i]$, we can construct a shorter path g in $\Gamma[\mu i, \lambda i]$, by setting $g(t) = \mathrm{Im}(f(t))\, i$. In order to complete the proof, we need only to show that if $g : [a, b] \to \mathbb{H}$ is any path in $\Gamma[\mu i, \lambda i]$ of the form $g(t) = y(t)i$, then

$$\mathrm{length}_{\mathbb{H}}(f_0) \leq \mathrm{length}_{\mathbb{H}}(g).$$

This follows immediately from Proposition 3.2. The image $g([a, b])$ of g is the hyperbolic line segment joining αi and βi, where $\alpha \leq \mu < \lambda \leq \beta$. Define $f_1 : [\alpha, \beta] \to \mathbb{H}$ by $f_1(t) = it$, and note that

$$\mathrm{length}_{\mathbb{H}}(f_0) = \ln\left[\frac{\lambda}{\mu}\right] \leq \ln\left[\frac{\beta}{\alpha}\right] = \mathrm{length}_{\mathbb{H}}(f_1).$$

Then, we can write $g = f_1 \circ (f_1^{-1} \circ g)$, where $f_1^{-1} \circ g : [a, b] \to [\alpha, \beta]$ is by construction a surjective function. By Proposition 3.2,

$$\mathrm{length}_{\mathbb{H}}(f_1) \leq \mathrm{length}_{\mathbb{H}}(g).$$

This completes the argument that

$$\mathrm{length}_{\mathbb{H}}(f_0) \leq \mathrm{length}_{\mathbb{H}}(f)$$

for every path f in $\Gamma[\mu i, \lambda i]$. That is, we have shown that

$$\mathrm{d}_{\mathbb{H}}(\mu i, \lambda i) = \mathrm{length}_{\mathbb{H}}(f_0) = \ln\left[\frac{\lambda}{\mu}\right].$$

Note that since we have written $g(t) = y(t)i$ and since $f_1(t) = it$, we have that $f_1^{-1} \circ g(t) = y(t)$, and so we have

$$\mathrm{length}_{\mathbb{H}}(g) = \mathrm{length}_{\mathbb{H}}(f_1)$$

if and only if either $y'(t) \geq 0$ for all t in $[a, b]$ or $y'(t) \leq 0$ for all t in $[a, b]$. That is, the only distance realizing paths in $\Gamma[\mu i, \lambda i]$ are those which are parametrizations of the hyperbolic line segment joining μi and λi.

Exercise 3.7

Consider the path $g : [-1, 1] \to \mathbb{H}$ given by

$$g(t) = (t^2 + 1)i.$$

Determine the image of g in $\mathbb{H}$ and calculate $\mathrm{length}_{\mathbb{H}}(g)$.

The transitivity of Möb($\mathbb{H}$) on the set of hyperbolic lines in $\mathbb{H}$ and the invariance of both hyperbolic lengths of paths in $\mathbb{H}$ and hyperbolic distances between pairs of points of $\mathbb{H}$ under the action of Möb($\mathbb{H}$) combine to yield that for any pair of points x and y in $\mathbb{H}$, there exists a distance realizing path in $\Gamma[x, y]$, namely a parametrization of the hyperbolic line segment joining x to y.

Namely, let ℓ be the hyperbolic line passing through x and y, and let γ be an element of Möb($\mathbb{H}$) taking ℓ to the positive imaginary axis I. Write $\gamma(x) = \mu i$ and $\gamma(y) = \lambda i$. Note that as before we can choose γ so that $\mu < \lambda$. If $\mu > \lambda$, replace γ with $K \circ \gamma$, where $K(z) = -\frac{1}{z}$.

We have just seen that the path $f_0 : [\mu, \lambda] \to \mathbb{H}$ given by $f_0(t) = ti$ is a distance realizing path in $\Gamma[\mu i, \lambda i]$. Since Möb($\mathbb{H}$) preserves hyperbolic lengths of paths, we have that

$$\text{length}_{\mathbb{H}}(\gamma^{-1} \circ f_0) = \text{length}_{\mathbb{H}}(f_0).$$

Since Möb($\mathbb{H}$) preserves hyperbolic distance, we have that

$$\mathrm{d}_{\mathbb{H}}(x, y) = \mathrm{d}_{\mathbb{H}}(\gamma^{-1}(\mu i), \gamma^{-1}(\lambda i)) = \mathrm{d}_{\mathbb{H}}(\mu i, \lambda i) = \text{length}_{\mathbb{H}}(f_0).$$

Combining these yields that

$$\text{length}_{\mathbb{H}}(\gamma^{-1} \circ f_0) = \mathrm{d}_{\mathbb{H}}(x, y),$$

and so $\gamma^{-1} \circ f_0$ is a distance realizing path in $\Gamma[x, y]$.

As mentioned at the beginning of this section, this also completes the proof that the second part of Condition 1 of the definition of a metric is satisfied by $\mathrm{d}_{\mathbb{H}}$. So, $(\mathbb{H}, \mathrm{d}_{\mathbb{H}})$ is a path metric space. This completes the proof of Theorem 3.10.

Exercise 3.8

Let S be the hyperbolic line segment between $2i$ and $10i$. For each $n \geq 2$, find the points that divide S into n segments of equal length.

Since we have that $\mathrm{d}_{\mathbb{H}}(x, y)$ is a metric on $\mathbb{H}$, the discussion in Note 3.9 yields that we now have notions of open and closed sets in $\mathbb{H}$, of convergent sequences of points of $\mathbb{H}$, and of continuous functions with domain and range $\mathbb{H}$.

We close this section by justifying why the boundary at infinity $\overline{\mathbb{R}} = \partial\mathbb{H}$ of $\mathbb{H}$ is called the boundary at infinity. Choose a point z on the boundary at infinity $\overline{\mathbb{R}}$ of $\mathbb{H}$, say $z = \infty$, and consider the hyperbolic ray ℓ determined by i and ∞.

Since ℓ can be expressed as the image of the path $f : [1, \infty) \to \mathbb{H}$ given by $f(t) = ti$, the distance between i and ∞ is equal to the length of f, namely the improper integral

$$\text{length}_{\mathbb{H}}(f) = \int_1^\infty \frac{1}{t}\, \mathrm{d}t,$$

which is infinite.

In particular, even though the points of $\overline{\mathbb{R}}$ form the topological boundary $\partial\mathbb{H}$ of $\mathbb{H}$ when we view $\mathbb{H}$ as a disc in $\overline{\mathbb{C}}$, the points of $\overline{\mathbb{R}}$ are infinitely far away from the points of $\mathbb{H}$ in terms of the hyperbolic metric on $\mathbb{H}$.

3.5 Formulae for Hyperbolic Distance in $\mathbb{H}$

The proof of Theorem 3.10 gives a method for calculating the hyperbolic distance between a pair of points in $\mathbb{H}$, at least in theory.

Given a pair of points x and y in $\mathbb{H}$, find or construct an element γ of Möb($\mathbb{H}$) so that $\gamma(x) = i\mu$ and $\gamma(y) = i\lambda$ both lie on the positive imaginary axis. Then, determine the values of μ and λ to find the hyperbolic distance

$$\mathrm{d}_{\mathbb{H}}(x, y) = \mathrm{d}_{\mathbb{H}}(\mu i, \lambda i) = \left| \ln \left[\frac{\lambda}{\mu} \right] \right|.$$

Note that here we use the absolute value, as we have made no assumption about whether $\lambda < \mu$ or $\mu < \lambda$.

For example, consider the two points $x = 2+i$ and $y = -3+i$. By Exercise 1.3, the hyperbolic line ℓ passing through x and y lies in the Euclidean circle with Euclidean centre $-\frac{1}{2}$ and Euclidean radius $\frac{\sqrt{29}}{2}$. In particular, the endpoints at infinity of ℓ are

$$p = \frac{-1+\sqrt{29}}{2} \text{ and } q = \frac{-1-\sqrt{29}}{2}.$$

Set $\gamma(z) = \frac{z-p}{z-q}$. The determinant of γ is $p - q > 0$, and so γ lies in Möb$^+$($\mathbb{H}$). Since by construction γ takes the endpoints at infinity of ℓ to the endpoints at infinity of the positive imaginary axis, namely 0 and ∞, we see that γ takes ℓ to the positive imaginary axis.

Calculating, we see that

$$\gamma(2+i) = \frac{2+i-p}{2+i-q} = \frac{p-q}{(2-q)^2+1}\, i$$

and

$$\gamma(-3+i) = \frac{-3+i-p}{-3+i-q} = \frac{p-q}{(3+q)^2+1}\, i.$$

In particular, we have that

$$\begin{aligned} d_{\mathbb{H}}(2+i, -3+i) = d_{\mathbb{H}}(\gamma(2+i), \gamma(-3+i)) &= \ln\left[\frac{(2-q)^2+1}{(3+q)^2+1}\right] \\ &= \ln\left[\frac{58+10\sqrt{29}}{58-10\sqrt{29}}\right]. \end{aligned}$$

As is demonstrated by this example, going through this procedure can be extremely tedious. It would be preferable to have an explicit and general formula of calculating hyperbolic distance. One way would be to repeat the procedure carried out in this example for a general pair of points z_1 and z_2.

Exercise 3.9

Let $z_1 = x_1 + y_1 i$ and $z_2 = x_2 + y_2 i$ be two points in $\mathbb{H}$ with $x_1 \neq x_2$. Derive a formula for $d_{\mathbb{H}}(z_1, z_2)$ in terms of x_1, y_1, x_2, and y_2 by constructing an element γ of $\text{Möb}(\mathbb{H})$ so that $\gamma(z_1)$ and $\gamma(z_2)$ both lie on the positive imaginary axis.

Exercise 3.10

Calculate the hyperbolic distance between each pair of the four points $A = i$, $B = 1+2i$, and $C = -1+2i$, and $D = 7i$.

A related formula for the hyperbolic distance $d_{\mathbb{H}}(z_1, z_2)$ between z_1 and z_2 in terms of their real and imaginary parts can be derived by making use of the fact that hyperbolic lines lie in Euclidean circles and Euclidean lines perpendicular to $\overline{\mathbb{R}}$. As above, write $z_1 = x_1 + y_1 i$ and $z_2 = x_2 + y_2 i$.

We can assume that $x_1 \neq x_2$, since in the case that $x_1 = x_2$, we have already seen that

$$d_{\mathbb{H}}(z_1, z_2) = \left|\ln\left[\frac{y_2}{y_1}\right]\right|.$$

Let c be the Euclidean centre and r the Euclidean radius of the Euclidean circle containing the hyperbolic line passing through z_1 and z_2. Suppose that $x_1 > x_2$, and let θ_k be the argument of z_k, taken in the range $[0, \pi)$ and as usual measured counter-clockwise from the positive real axis.

Consider the path $f : [\theta_1, \theta_2] \to \mathbb{H}$ given by $f(t) = c + re^{it}$. The image of f is the hyperbolic line segment between z_1 and z_2, and so $d_{\mathbb{H}}(z_1, z_2) = \text{length}_{\mathbb{H}}(f)$.

Since $\text{Im}(f(t)) = r\sin(t)$ and $|f'(t)| = |rie^{it}| = r$, we have that

$$d_{\mathbb{H}}(z_1, z_2) = \text{length}_{\mathbb{H}}(f) = \int_{\theta_1}^{\theta_2} \frac{1}{\sin(t)} dt = \ln\left|\frac{\csc(\theta_2) - \cot(\theta_2)}{\csc(\theta_1) - \cot(\theta_1)}\right|.$$

In order to rewrite this expression in terms of x_1, x_2, y_1, and y_2, it is possible but not necessary to express the θ_k in terms of the x_k and y_k. We might also express $\csc(\theta_k)$ and $\cot(\theta_k)$ in terms of the x_k and y_k, and c and r.

Note that θ_k is the angle of the right triangle with opposite side y_k, adjacent side $x_k - c$, and hypotenuse r. So, we have that

$$\csc(\theta_k) = \frac{r}{y_k} \text{ and } \cot(\theta_k) = \frac{x_k - c}{y_k}.$$

This gives that

$$|\csc(\theta_k) - \cot(\theta_k)| = \left|\frac{r + c - x_k}{y_k}\right|,$$

and so

$$\mathrm{d}_{\mathbb{H}}(z_1, z_2) = \mathrm{length}_{\mathbb{H}}(f) = \ln\left|\frac{\csc(\theta_2) - \cot(\theta_2)}{\csc(\theta_1) - \cot(\theta_1)}\right| = \ln\left|\frac{(x_1 - c - r)y_2}{y_1(x_2 - c - r)}\right|.$$

Note that if instead we have that $x_2 < x_1$ and we go through this calculation, we get that

$$\mathrm{d}_{\mathbb{H}}(z_1, z_2) = \mathrm{length}_{\mathbb{H}}(f) = \ln\left|\frac{\csc(\theta_1) - \cot(\theta_1)}{\csc(\theta_2) - \cot(\theta_2)}\right| = \ln\left|\frac{y_1(x_2 - c - r)}{(x_1 - c - r)y_2}\right|,$$

which differs from $\ln\left|\frac{(x_1-c-r)y_2}{y_1(x_2-c-r)}\right|$ by a factor of -1.

So, if we make no assumption of the relationship between x_1 and x_2, we obtain the formula

$$\mathrm{d}_{\mathbb{H}}(z_1, z_2) = \left|\ln\left|\frac{(x_1 - c - r)y_2}{y_1(x_2 - c - r)}\right|\right|$$

for the hyperbolic distance between z_1 and z_2.

If we wish to express this formula solely in terms of the x_k and y_k, we may recall the result of Exercise 1.3, in which we gave expressions for c and r in terms of the x_k and y_k. Unfortunately, the resulting expression does not simplify very much, and so we do not give it explicitly here.

Though it can be unwieldy, we can sometimes make explicit use of this formula. For example, we can determine whether or not there exists a positive real number s so that

$$\mathrm{d}_{\mathbb{H}}(-s + i, i) = \mathrm{d}_{\mathbb{H}}(i, s + i) = \mathrm{d}_{\mathbb{H}}(-s + i, s + i).$$

Since $-s + i$ and $s + i$ lie on the Euclidean circle with Euclidean centre $c = 0$ and Euclidean radius $r = \sqrt{1 + s^2}$, we have that

$$\mathrm{d}_{\mathbb{H}}(-s + i, s + i) = \ln\left[\frac{\sqrt{s^2 + 1} + s}{\sqrt{s^2 + 1} - s}\right].$$

Since $s+i$ and i lie on the Euclidean circle with Euclidean centre $c = \frac{s}{2}$ and Euclidean radius $r = \frac{1}{2}\sqrt{4+s^2}$, we have that

$$\mathrm{d}_{\mathbb{H}}(s+i, i) = \ln \left[\frac{\sqrt{s^2+4}+s}{\sqrt{s^2+4}-s} \right].$$

Since there is no value of s for which

$$\ln \left[\frac{\sqrt{s^2+1}+s}{\sqrt{s^2+1}-s} \right] = \ln \left[\frac{\sqrt{s^2+4}+s}{\sqrt{s^2+4}-s} \right],$$

no such value of s exists.

Now that we understand hyperbolic distance, and specifically now that we have a notion of hyperbolic distance that is invariant under the action of Möb($\mathbb{H}$), we are able to see the obstruction to Möb($\mathbb{H}$) acting transitively on pairs of distinct points of $\mathbb{H}$.

Exercise 3.11

Given two pairs (z_1, z_2) and (w_1, w_2) of distinct points of $\mathbb{H}$, prove that there exists an element q of Möb($\mathbb{H}$) taking one to the other if and only if $\mathrm{d}_{\mathbb{H}}(z_1, z_2) = \mathrm{d}_{\mathbb{H}}(w_1, w_2)$.

3.6 Isometries

In general, an *isometry* of a metric space (X, d) is a homeomorphism f of X that preserves distance. That is, an isometry of (X, d) is a homeomorphism f of X for which

$$\mathrm{d}(x, y) = \mathrm{d}(f(x), f(y))$$

for every pair x and y of points of X. In fact, as is demonstrated in the following exercise, this definition of an isometry is partially redundant.

Exercise 3.12

Let $f : X \to X$ be any function that preserves distance. Prove that f is injective and continuous.

In general, we cannot conclude that a distance preserving function $f : X \to X$ is a homeomorphism. To illustrate one thing that can go wrong, consider the function $\mathrm{e} : \mathbb{Z} \times \mathbb{Z} \to \mathbb{R}$ defined by

$$\mathrm{e}(n, m) = \begin{cases} 0 & \text{if } m = n, \text{ and} \\ 1 & \text{if } m \neq n. \end{cases}$$

This gives a metric on $\mathbb{Z}$, which is admittedly very different from the usual metric on $\mathbb{Z}$. The function $f : \mathbb{Z} \to \mathbb{Z}$ defined by $f(m) = 2m$ is distance preserving but is not surjective, and hence is not a homeomorphism.

It is true, though, that a distance preserving function $f : X \to X$ is a homeomorphism onto its image $f(X)$, since f is a bijection when considered as a function $f : X \to f(X)$. For each pair of points z and w of $f(X)$, we have that

$$\mathrm{d}(z, w) = \mathrm{d}(f(f^{-1}(z)), f(f^{-1}(w))) = \mathrm{d}(f^{-1}(z), f^{-1}(w)).$$

Hence, $f^{-1} : f(X) \to X$ is also a distance preserving function, and so is continuous by Exercise 3.12.

Exercise 3.13

Prove that the function $f : \mathbb{C} \to \mathbb{C}$ given by $f(z) = az$ is an isometry of $(\mathbb{C}, \mathrm{n})$ if and only if $|a| = 1$. Here, $\mathrm{n}(z, w) = |z - w|$, as in Section 3.3.

The only metric space we have studied so far in any detail is the upper half-plane model $(\mathbb{H}, \mathrm{d}_{\mathbb{H}})$ of the hyperbolic plane. Define a *hyperbolic isometry* to be an isometry of $(\mathbb{H}, \mathrm{d}_{\mathbb{H}})$. In this section, we characterize the hyperbolic isometries as precisely the elements of $\text{Möb}(\mathbb{H})$.

Since the inverse of an invertible distance preserving map is necessarily distance preserving, and since the composition of two distance preserving maps is distance preserving, the set of all isometries of a metric space is a group. Let $\text{Isom}(\mathbb{H})$ denote the *group of isometries* of $(\mathbb{H}, \mathrm{d}_{\mathbb{H}})$.

Theorem 3.12

$\text{Isom}(\mathbb{H}) = \text{Möb}(\mathbb{H})$.

By our construction of the hyperbolic metric $\mathrm{d}_{\mathbb{H}}$ on $\mathbb{H}$, specifically Proposition 3.11, we have that every element of $\text{Möb}(\mathbb{H})$ is a hyperbolic isometry, and so $\text{Möb}(\mathbb{H}) \subset \text{Isom}(\mathbb{H})$.

We begin the proof of the opposite inclusion with the observation that hyperbolic line segments can be characterized purely in terms of hyperbolic distance.

Proposition 3.13

Let x, y, and z be distinct points in $\mathbb{H}$. Then,

$$d_{\mathbb{H}}(x, y) + d_{\mathbb{H}}(y, z) = d_{\mathbb{H}}(x, z)$$

if and only if y is contained in the hyperbolic line segment ℓ_{xz} joining x to z.

To begin the proof of Proposition 3.13, let m be an element of Möb($\mathbb{H}$) for which $m(x) = i$ and $m(z) = \alpha\, i$ for some $\alpha > 1$.

[To see that such an m exists, first take an element γ of Möb($\mathbb{H}$) taking the hyperbolic line passing through x and z to the positive imaginary axis. Write $\gamma(x) = \mu i$ and $\gamma(z) = \lambda i$. If necessary, replace γ by $K \circ \gamma$, where $K(z) = -\frac{1}{z}$, to ensure that $\mu < \lambda$. We then compose γ with $\varphi(z) = \frac{1}{\mu} z$, so that

$$\varphi \circ \gamma(x) = i \text{ and } \varphi \circ \gamma(z) = \alpha i,$$

where $\alpha > 1$. Note that

$$d_{\mathbb{H}}(x, z) = d_{\mathbb{H}}(i, \alpha i) = \ln(\alpha),$$

and so

$$\alpha = e^{d_{\mathbb{H}}(x,z)}.]$$

Write $m(y) = a + bi$. There are several cases to consider.

Suppose that y lies on the hyperbolic line segment ℓ_{xz} joining x to z. Then, $m(y)$ lies on the hyperbolic line segment joining $m(x) = i$ to $m(z) = \alpha i$. In particular, $a = 0$ and $1 \le b \le \alpha$, and so

$$d_{\mathbb{H}}(x, y) = d_{\mathbb{H}}(i, bi) = \ln(b)$$

and

$$d_{\mathbb{H}}(y, z) = d_{\mathbb{H}}(bi, \alpha i) = \ln\left[\frac{\alpha}{b}\right] = d_{\mathbb{H}}(x, z) - \ln(b).$$

Hence, $d_{\mathbb{H}}(x, z) = d_{\mathbb{H}}(x, y) + d_{\mathbb{H}}(y, z)$.

Suppose now that y does not lie on the hyperbolic line segment ℓ_{xz} joining x to z. There are two cases, namely that $m(y)$ lies on the positive imaginary axis, so that $a = 0$, and that $m(y)$ does not lie on the positive imaginary axis, so that $a \neq 0$.

If $a = 0$, then $m(y) = bi$, where either $0 < b < 1$ or $\alpha < b$.

If $0 < b < 1$, then

$$d_{\mathbb{H}}(x, y) = -\ln(b) \text{ and } d_{\mathbb{H}}(y, z) = \ln\left[\frac{\alpha}{b}\right] = d_{\mathbb{H}}(x, z) - \ln(b).$$

Since $\ln(b) < 0$, we have that

$$\mathrm{d_{\mathbb{H}}}(x,y) + \mathrm{d_{\mathbb{H}}}(y,z) = \mathrm{d_{\mathbb{H}}}(x,z) - 2\ln(b) > \mathrm{d_{\mathbb{H}}}(x,z).$$

If $b > \alpha$, then

$$\mathrm{d_{\mathbb{H}}}(x,y) = \ln(b) \text{ and } \mathrm{d_{\mathbb{H}}}(y,z) = \ln\left[\frac{b}{\alpha}\right] = \ln(b) - \mathrm{d_{\mathbb{H}}}(x,z).$$

Since $\ln(b) > \mathrm{d_{\mathbb{H}}}(x,z)$, we have that

$$\mathrm{d_{\mathbb{H}}}(x,y) + \mathrm{d_{\mathbb{H}}}(y,z) = 2\ln(b) - \mathrm{d_{\mathbb{H}}}(x,z) > \mathrm{d_{\mathbb{H}}}(x,z).$$

If $a \neq 0$, we begin with the observation that

$$\mathrm{d_{\mathbb{H}}}(i, bi) < \mathrm{d_{\mathbb{H}}}(i, a+bi) = \mathrm{d_{\mathbb{H}}}(x,y).$$

This follows from the argument given in Section 3.4.

Specifically, let $f : [\alpha,\beta] \to \mathbb{H}$ be a distance realizing path between $i = f(\alpha)$ and $a + bi = f(\beta)$. Note that the path $g : [\alpha,\beta] \to \mathbb{H}$ given by $g(t) = \mathrm{Im}(f(t))\, i$ satisfies $g(\alpha) = i$, $g(\beta) = bi$, and $\mathrm{length_{\mathbb{H}}}(g) < \mathrm{length_{\mathbb{H}}}(f)$.

Similarly, we have that

$$\mathrm{d_{\mathbb{H}}}(bi, \alpha i) < \mathrm{d_{\mathbb{H}}}(a+bi, \alpha i) = \mathrm{d_{\mathbb{H}}}(y,z).$$

If $1 \leq b \leq \alpha$, then

$$\mathrm{d_{\mathbb{H}}}(x,z) = \mathrm{d_{\mathbb{H}}}(i,\alpha i) = \mathrm{d_{\mathbb{H}}}(i,bi) + \mathrm{d_{\mathbb{H}}}(bi,\alpha i) < \mathrm{d_{\mathbb{H}}}(x,y) + \mathrm{d_{\mathbb{H}}}(y,z).$$

If b does not lie in $[1,\alpha]$, then again we have two cases, namely that $0 < b < 1$ and that $\alpha < b$.

Making use of the calculations of the previous few paragraphs, in the case that $0 < b < 1$ we have

$$\mathrm{d_{\mathbb{H}}}(x,z) < \mathrm{d_{\mathbb{H}}}(x,z) - 2\ln(b) = \mathrm{d_{\mathbb{H}}}(i,bi) + \mathrm{d_{\mathbb{H}}}(bi,\alpha i) < \mathrm{d_{\mathbb{H}}}(x,y) + \mathrm{d_{\mathbb{H}}}(y,z),$$

while in the case that $b > \alpha$ we have

$$\mathrm{d_{\mathbb{H}}}(x,z) < 2\ln(b) - \mathrm{d_{\mathbb{H}}}(x,z) = \mathrm{d_{\mathbb{H}}}(i,bi) + \mathrm{d_{\mathbb{H}}}(bi,\alpha i) < \mathrm{d_{\mathbb{H}}}(x,y) + \mathrm{d_{\mathbb{H}}}(y,z).$$

This completes the proof of Proposition 3.13.

Exercise 3.14

Prove that every hyperbolic isometry of $\mathbb{H}$ takes hyperbolic lines to hyperbolic lines.

Let f be a hyperbolic isometry, and recall that we are proving that f is an element of $\text{Möb}(\mathbb{H})$. For each pair of points p and q of $\mathbb{H}$, let ℓ_{pq} denote the hyperbolic line segment joining p to q. With this notation, Proposition 3.13 can be rephrased as saying that $\ell_{f(p)f(q)} = f(\ell_{pq})$.

Let ℓ be the perpendicular bisector of the hyperbolic line segment ℓ_{pq}, which is defined to be the hyperbolic line

$$\ell = \{z \in \mathbb{H} \mid \mathrm{d}_{\mathbb{H}}(p,z) = \mathrm{d}_{\mathbb{H}}(q,z)\}.$$

Since ℓ is defined in terms of hyperbolic distance, we have that $f(\ell)$ is the perpendicular bisector of $f(\ell_{pq}) = \ell_{f(p)f(q)}$.

We now normalize the hyperbolic isometry f. Pick a pair of points x and y on the positive imaginary axis I in $\mathbb{H}$, and let H be one of the half-planes in $\mathbb{H}$ determined by I.

By Exercise 3.11, there exists an element γ of $\text{Möb}(\mathbb{H})$ that satisfies $\gamma(f(x)) = x$ and $\gamma(f(y)) = y$, since $\mathrm{d}_{\mathbb{H}}(x,y) = \mathrm{d}_{\mathbb{H}}(f(x), f(y))$. In particular, we see that $\gamma \circ f$ fixes both x and y, and so $\gamma \circ f$ takes I to I. If necessary, replace γ by the composition $B \circ \gamma$ with the reflection $B(z) = -\overline{z}$ in I to obtain an element γ of $\text{Möb}(\mathbb{H})$ so that $\gamma \circ f$ takes I to I and also takes H to H.

Let z be any point on I. Since z is uniquely determined by the two hyperbolic distances $\mathrm{d}_{\mathbb{H}}(x,z)$ and $\mathrm{d}_{\mathbb{H}}(y,z)$ and since both hyperbolic distances are preserved by $\gamma \circ f$, we have that $\gamma \circ f$ fixes every point z of I.

Exercise 3.15

Let $x = \lambda i$ and $z = \mu i$ be two distinct points on the positive imaginary axis I. Let y be any point on I. Show that y is uniquely determined by the two hyperbolic distances $\mathrm{d}_{\mathbb{H}}(x,y)$ and $\mathrm{d}_{\mathbb{H}}(y,z)$.

Now, let w be any point in $\mathbb{H}$ that does not lie on I, and let ℓ be the hyperbolic line through w that is perpendicular to I. Explicitly, we can describe ℓ as the hyperbolic line contained in the Euclidean circle with Euclidean centre 0 and Euclidean radius $|w|$. Let z be the point of intersection of ℓ and I.

At this point, we know several facts about ℓ. Since ℓ is the perpendicular bisector of some hyperbolic line segment in I and since $\gamma \circ f$ fixes every point of I, we have that $\gamma \circ f(\ell) = \ell$.

Since $\gamma \circ f$ fixes z, since $\mathrm{d}_{\mathbb{H}}(z,w) = \mathrm{d}_{\mathbb{H}}(\gamma \circ f(z), \gamma \circ f(w)) = \mathrm{d}_{\mathbb{H}}(z, \gamma \circ f(w))$, and since $\gamma \circ f$ preserves the two half-planes determined by I, we have that $\gamma \circ f$ fixes w.

Since $\gamma \circ f$ fixes every point of $\mathbb{H}$, we have that $\gamma \circ f$ is the identity. In particular, we have that $f = \gamma^{-1}$, and so f is an element of Möb($\mathbb{H}$). This completes the proof of Theorem 3.12.

Note that in this proof of Theorem 3.12, we introduced a new system of coordinates on $\mathbb{H}$, different from the standard coordinates coming from $\mathbb{C}$. This new set of coordinates comes from locating a point in $\mathbb{H}$ relative to the positive imaginary axis I and all the hyperbolic lines perpendicular to I.

Explicitly, let w be any point in $\mathbb{H}$. We first note that w lies on the Euclidean circle with Euclidean centre 0 and Euclidean radius $|w|$. This Euclidean circle contains the hyperbolic line ℓ_w through w that is perpendicular to I. Let $Z(w)$ denote the point of intersection of I and ℓ_w.

We can uniquely locate w on ℓ_w by considering its *signed hyperbolic distance* from the positive imaginary axis, which is the number

$$\mathrm{sign}(w)\, \mathrm{d}_{\mathbb{H}}(Z(w), w),$$

where $\mathrm{sign}(w) = 1$ if $\mathrm{Re}(w) > 0$, where $\mathrm{sign}(w) = -1$ if $\mathrm{Re}(w) < 0$, and where $\mathrm{sign}(w) = 0$ if $\mathrm{Re}(w) = 0$.

To w we then associate the coordinates

$$(\log(|w|), \mathrm{sign}(w)\, \mathrm{d}_{\mathbb{H}}(Z(w), w)) .$$

For an illustration of these coordinates, see Fig. 3.1.

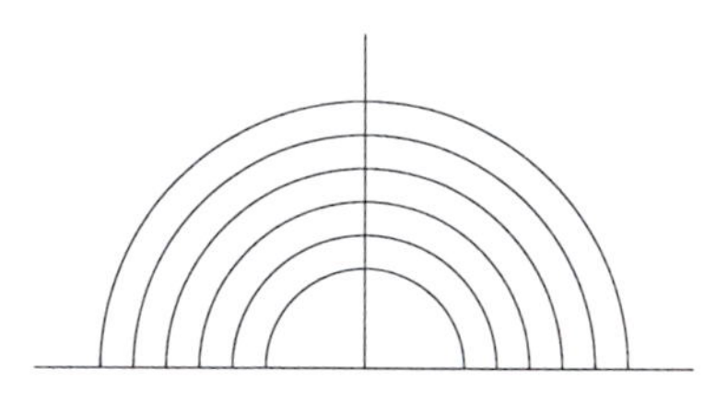

Figure 3.1: New coordinates on $\mathbb{H}$.

For example, consider the point $z = 1 + 2i$. The hyperbolic line ℓ_z passing through z and perpendicular to I lies in the Euclidean circle with Euclidean centre 0 and Euclidean radius $\sqrt{5}$. The point of intersection of I and ℓ_z is $Z(z) = \sqrt{5}i$. The signed hyperbolic distance from z to I is then

$$\mathrm{d}_{\mathbb{H}}(\sqrt{5}i, 1 + 2i) = \frac{2}{\sqrt{5} - 1},$$

since $\operatorname{sign}(z) = 1$. So, in these new coordinates, $z = 1 + 2i$ corresponds to the point

$$\left(\log(\sqrt{5}), \frac{2}{\sqrt{5}-1}\right).$$

Exercise 3.16

Express the action of $m(z) = 2z$ on $\mathbb{H}$ in terms of these coordinates.

3.7 Metric Properties of $(\mathbb{H}, \mathrm{d}_{\mathbb{H}})$

In this section, we investigate some properties of the hyperbolic metric on $\mathbb{H}$.

In much the same way that we can define the hyperbolic distance between a pair of points, there is a notion of the hyperbolic distance between a pair X and Y of subsets of $\mathbb{H}$, namely

$$\mathrm{d}_{\mathbb{H}}(X, Y) = \inf\{\mathrm{d}_{\mathbb{H}}(x, y) \,:\, x \in X,\ y \in Y\}.$$

As we will see later in this section, there exist disjoint sets X and Y in $\mathbb{H}$ for which $\mathrm{d}_{\mathbb{H}}(X, Y) = 0$, and so this does not give a metric on the set of subsets of $\mathbb{H}$.

In general, calculating this infimum can be very difficult. We spend some of this section exploring in some detail the case that one or both of X and Y are hyperbolic lines. There is one general fact about this distance between sets that will prove to be very useful.

We first need to make a definition.

Definition 3.14

A subset X of $\mathbb{H}$ is *bounded* if there exists some $C > 0$ so that X is contained in the open hyperbolic disc

$$U_C(i) = \{z \in \mathbb{H} \mid \mathrm{d}_{\mathbb{H}}(z, i) < C\}.$$

A subset X of $\mathbb{H}$ is *compact* if X is closed and bounded.

One easy example of a compact subset of $\mathbb{H}$ is any set containing a finite number of points $X = \{x_1, \ldots, x_n\}$. For any z in $\mathbb{H} - X$, set

$$\varepsilon = \inf\{\mathrm{d}_{\mathbb{H}}(z, x_1), \ldots, \mathrm{d}_{\mathbb{H}}(z, x_n)\}.$$

Then, $\varepsilon > 0$ and $U_\varepsilon(z)$ is contained in $\mathbb{H} - X$, so that $\mathbb{H} - X$ is open and so X is closed. Also, if we set

$$C = \sup\{\mathrm{d}_{\mathbb{H}}(i, x_1), \ldots, \mathrm{d}_{\mathbb{H}}(i, x_n)\},$$

then X is contained in $U_{2C}(i)$, and so X is bounded.

Though we do not prove it here, a basic property of compact sets is that if X is a compact subset of $\mathbb{H}$ and if $\{x_n\}$ is a sequence of points of X, then there is a subsequence $\{x_{n_k}\}$ of $\{x_n\}$ so that $\{x_{n_k}\}$ converges to a point x of X. In words, a sequence of points in a compact set X contains a convergent subsequence.

Exercise 3.17

Let X be a compact subset of $\mathbb{H}$ and let Y be any subset of $\mathbb{H}$. Prove that $\mathrm{d}_{\mathbb{H}}(X, Y) > 0$ if and only if X and Y have disjoint closures.

Though this notion of hyperbolic distance between sets does not give a metric on the set of subsets of $\mathbb{H}$, it does give one way of measuring when two subsets of $\mathbb{H}$ are close. A particularly interesting application of this notion is to pairs of hyperbolic lines and hyperbolic rays.

Recall that there are two different types of parallelism for pairs of hyperbolic lines. There are pairs of hyperbolic lines that are disjoint in $\mathbb{H}$ but for which the circles in $\overline{\mathbb{C}}$ containing them are not disjoint, and there are pairs of hyperbolic lines that are disjoint in $\mathbb{H}$ and for which the circles in $\overline{\mathbb{C}}$ containing them are also disjoint. We refer to the former hyperbolic lines as *parallel* and the latter hyperbolic lines as *ultraparallel.*

We saw in Section 1.3 that we can distinguish these two cases by examining the endpoints at infinity of the two hyperbolic lines. Now that we have a means of measuring hyperbolic distance, we can distinguish these two cases intrinsically as well.

Let ℓ_0 and ℓ_1 be parallel hyperbolic lines that share an endpoint at infinity at the point x of $\overline{\mathbb{R}}$. Let y_k be the other endpoint at infinity of ℓ_k. Since by Proposition 2.30 we have that Möb($\mathbb{H}$) acts triply transitively on $\overline{\mathbb{R}}$, we may assume that $x = \infty$, that $y_0 = 0$, and that $y_1 = 1$.

We now calculate. Each point of ℓ_0 has the form λi for some $\lambda > 0$, and each point of ℓ_1 has the form $1 + \lambda i$ for some $\lambda > 0$.

The path $f : [0, 1] \to \mathbb{H}$ given by $f(t) = t + \lambda i$ parametrizes the horizontal Euclidean line segment joining λi and $1 + \lambda i$, and so

$$\mathrm{d}_{\mathbb{H}}(\ell_0, \ell_1) \leq \mathrm{d}_{\mathbb{H}}(\lambda i, 1 + \lambda i) \leq \mathrm{length}_{\mathbb{H}}(f) = \int_0^1 \frac{1}{\lambda}\, \mathrm{d}t = \frac{1}{\lambda}$$

for every $\lambda > 0$. Letting λ tend to ∞, we see that

$$d_{\mathbb{H}}(\ell_0, \ell_1) = 0$$

for two parallel hyperbolic lines ℓ_0 and ℓ_1 that share an endpoint at infinity.

Suppose on the other hand that ℓ_0 and ℓ_1 are ultraparallel hyperbolic lines.

Proposition 3.15

Let ℓ_0 and ℓ_1 be ultraparallel hyperbolic lines in $\mathbb{H}$. Then, $d_{\mathbb{H}}(\ell_0, \ell_1) > 0$.

Again by making use of the triple transitivity of Möb($\mathbb{H}$) on $\overline{\mathbb{R}}$, we may assume that the endpoints at infinity of ℓ_0 are 0 and ∞, and that the endpoints at infinity of ℓ_1 are 1 and $x > 1$. We wish to calculate the hyperbolic distance $d_{\mathbb{H}}(\ell_0, \ell_1)$ between ℓ_0 and ℓ_1.

We make use of the following fact.

Exercise 3.18

Let ℓ be a hyperbolic line and let p be a point of $\mathbb{H}$ not on ℓ. Prove that there exists a unique point z on ℓ so that the hyperbolic line segment through z and p is perpendicular to ℓ, and so that

$$d_{\mathbb{H}}(p, \ell) = d_{\mathbb{H}}(p, z).$$

For each $1 < r < x$, let c_r be the hyperbolic line contained in the Euclidean circle with Euclidean centre 0 and Euclidean radius r, so that c_r is perpendicular to ℓ_0 for every r. Note that we could define c_r for every $r > 0$, but c_r intersects ℓ_1 only for $1 < r < x$. Write the point of intersection of c_r and ℓ_1 as $re^{i\theta}$, where $0 < \theta < \frac{\pi}{2}$.

We can determine θ by considering the Euclidean triangle with vertices 0, $\frac{1}{2}(x+1)$ (the Euclidean centre of the Euclidean circle containing ℓ_1), and $re^{i\theta}$. See Fig. 3.2.

The Euclidean lengths of the two sides of this Euclidean triangle adjacent to the vertex 0, which has angle θ, are r and $\frac{1}{2}(x+1)$, and the length of the opposite side is $\frac{1}{2}(x-1)$. Calculating, we see that

$$\left[\frac{1}{2}(x-1)\right]^2 = \left[\frac{1}{2}(x+1)\right]^2 + r^2 - 2r\left[\frac{1}{2}(x+1)\right]\cos(\theta)$$

by the law of cosines.

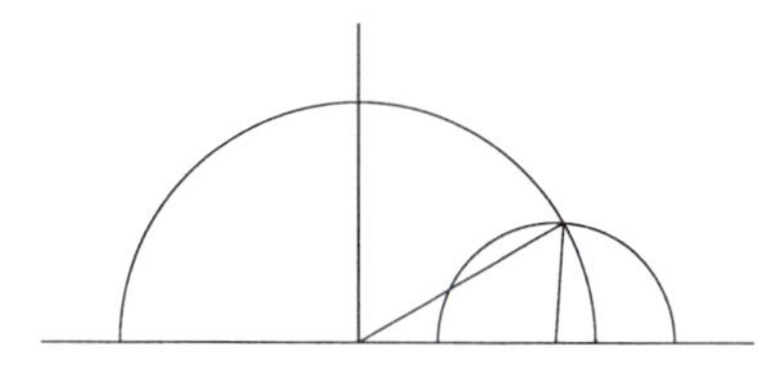

Figure 3.2: The Euclidean triangle in $\mathbb{H}$ with vertices 0, $\frac{1}{2}(x+1)$, and $re^{i\theta}$.

Simplifying, we see that this is equivalent to

$$x + r^2 = r(x+1)\cos(\theta),$$

and so

$$\cos(\theta) = \frac{x+r^2}{r(x+1)}$$

and

$$\sin(\theta) = \sqrt{1-\cos^2(\theta)} = \frac{\sqrt{(r^2-1)(x^2-r^2)}}{r(x+1)}.$$

The hyperbolic distance between ri and $re^{i\theta}$ for this value of θ is the length of the hyperbolic line segment joining ri and $re^{i\theta}$. Parametrizing this hyperbolic line segment by $f(t) = re^{it}$ for $\theta \le t \le \frac{\pi}{2}$, we calculate that

$$\text{length}_{\mathbb{H}}(f) = \int_\theta^{\frac{\pi}{2}} \frac{1}{\sin(t)}\,\mathrm{d}t = -\ln|\csc(\theta)-\cot(\theta)| = \frac{1}{2}\ln\left[\frac{(r+1)(x+r)}{(r-1)(x-r)}\right].$$

Since c_r is perpendicular to ℓ_0, we know by Exercise 3.18 that

$$\mathrm{d}_{\mathbb{H}}(re^{i\theta}, ri) = \mathrm{d}_{\mathbb{H}}(re^{i\theta}, \ell_0).$$

In particular, the hyperbolic distance between the two hyperbolic lines ℓ_0 and ℓ_1 is the minimum hyperbolic distance between $re^{i\theta}$ and ri as r varies over the interval $(1, x)$.

The hyperbolic distance between $re^{i\theta}$ and ri is minimized when

$$\frac{\mathrm{d}}{\mathrm{d}r}\ln\left[\frac{(r+1)(x+r)}{(r-1)(x-r)}\right] = \frac{2(r^2-x)(x+1)}{(r+1)(x+r)(r-1)(x-r)} = 0.$$

Since $r > 0$, this can only occur when $r = \sqrt{x}$.

Hence, the hyperbolic distance between the two hyperbolic lines ℓ_0 and ℓ_1 is

$$\mathrm{d}_{\mathbb{H}}(\ell_0, \ell_1) = \frac{1}{2}\ln\left[\frac{(\sqrt{x}+1)(x+\sqrt{x})}{(\sqrt{x}-1)(x-\sqrt{x})}\right] = \ln\left[\frac{\sqrt{x}+1}{\sqrt{x}-1}\right],$$

which is positive since $x > 1$. This completes the proof of Proposition 3.15.

One consequence of the proof of Proposition 3.15 is that it also shows that there exists a unique common perpendicular for any pair of ultraparallel hyperbolic lines.

Proposition 3.16

Let ℓ_0 and ℓ_1 be two ultraparallel hyperbolic lines. Then, there exists a unique hyperbolic line ℓ that is perpendicular to both ℓ_0 and ℓ_1.

We use the same notation and normalizations as in the proof of Proposition 3.15. We know that c_r is perpendicular to ℓ_0 for all values of r by construction.

To determine for which values of r we have that c_r is perpendicular to ℓ_1, we apply the Pythagorean theorem to the Euclidean triangle with vertices 0, $\frac{1}{2}(x+1)$, and $re^{i\theta}$. The angle between c_r and ℓ_1 is $\frac{\pi}{2}$ if and only if

$$\left[\frac{1}{2}(x+1)\right]^2 = \left[\frac{1}{2}(x-1)\right]^2 + r^2,$$

which occurs if and only if $r = \sqrt{x}$. This completes the proof of Proposition 3.16.

Exercise 3.19

Let I be the positive imaginary axis in $\mathbb{H}$. For a positive real number $\varepsilon > 0$, let W_ε be the set of points in $\mathbb{H}$ whose hyperbolic distance from I is equal to ε. Prove that W_ε is the union of two Euclidean rays from 0 that make equal angle θ with I. Relate θ to ε.

Exercise 3.20

Prove that if ℓ_0 and ℓ_1 are hyperbolic lines that share an endpoint at infinity, then there does not exist a hyperbolic line perpendicular to both ℓ_0 and ℓ_1.

Exercise 3.21

Let ℓ_0 and ℓ_1 be parallel hyperbolic lines in $\mathbb{H}$. Label the endpoints at infinity of ℓ_0 as z_0 and z_1, and the endpoints at infinity of ℓ_1 as w_0 and w_1, so that they occur in the order z_0, w_0, w_1, z_1 moving counter-clockwise around $\overline{\mathbb{R}}$. Prove that

$$\tanh^2\left[\frac{1}{2}\mathrm{d}_{\mathbb{H}}(\ell_0,\ell_1)\right] = \frac{1}{1-[z_0,w_0;w_1,z_1]}.$$

Though we will not explore it in detail, we do note here that this notion of distance between sets can be used to give a description of the boundary at infinity of $\mathbb{H}$ that is intrinsic to $\mathbb{H}$ and that does not make use of how $\mathbb{H}$ sits as a subset of $\overline{\mathbb{C}}$.

Let $\mathcal{R}$ be the set of all hyperbolic rays in $\mathbb{H}$. For each ray R in $\mathcal{R}$, let $\text{sub}(R)$ be the set of all the subrays of R, which are the hyperbolic rays contained in R. Given any two rays R_1 and R_2 in $\mathcal{R}$, say that $R_1 \sim R_2$ if and only if

$$\sup\{\mathrm{d}_{\mathbb{H}}(R_1^0, R_2^0) \mid R_1^0 \in \text{sub}(R_1), R_2^0 \in \text{sub}(R_2)\} = 0.$$

Note that if two non-equal hyperbolic rays R_1 and R_2 have the same initial point in $\mathbb{H}$, then this supremum is infinite, and so $R_1 \not\sim R_2$. In fact, for any two hyperbolic rays, this supremum is either 0 or infinite, and is 0 if and only if the two rays have the same endpoint at infinity.

This gives a way of identifying the boundary at infinity $\overline{\mathbb{R}}$ of $\mathbb{H}$ with equivalence classes in $\mathcal{R}$. Morever, since elements of Möb($\mathbb{H}$) take hyperbolic rays to hyperbolic rays and preserve hyperbolic distance, we see that Möb($\mathbb{H}$) preserves the equivalence relation, and so we get an action of Möb($\mathbb{H}$) on $\mathcal{R}/\sim$.

Exercise 3.22

Let ℓ be a hyperbolic line in $\mathbb{H}$, and let p be a point in $\mathbb{H}$ not on ℓ. Determine the proportion of the hyperbolic rays from p that intersect ℓ.

4
Other Models of the Hyperbolic Plane

Up to this point, we have focused our attention exclusively on the upper half-plane model $\mathbb{H}$ of the hyperbolic plane, but there are many other useful models. We explore in this chapter a second particular model, the *Poincaré disc model* $\mathbb{D}$, of the hyperbolic plane, which we construct starting from the upper half-plane model. We go on to show that the construction used for the Poincaré disc is but one instance of a *general construction* for producing planar models of the hyperbolic plane.

4.1 The Poincaré Disc Model

Up to this point, we have focused our attention on developing the upper half-plane model $\mathbb{H}$ of the hyperbolic plane and studying its properties. There are a number of other models of the hyperbolic plane. One of the most useful of these other models, at least for our purposes, is the *Poincaré disc* model $\mathbb{D}$.

There are a number of ways we could develop this, and other, models of the hyperbolic plane. One very inefficient way is to retrace all the steps we undertook to develop the upper half-plane model. Another way is to make use of what we have done in developing the upper half-plane model, and find a way of transferring this work to the other model. We take the latter approach.

The underlying space of the Poincaré model of the hyperbolic plane is the open unit disc

$$\mathbb{D} = \{z \in \mathbb{C} \mid |z| < 1\}$$

in the complex plane $\mathbb{C}$. Since $\mathbb{H}$ and $\mathbb{D}$ are both discs in the Riemann sphere $\overline{\mathbb{C}}$, we know from Theorem 2.11 that there exists an element m of Möb taking $\mathbb{D}$ to $\mathbb{H}$. In fact, in Exercise 2.10, you constructed an explicit element of Möb taking $\mathbb{D}$ to $\mathbb{H}$.

We now use m to transport hyperbolic geometry from $\mathbb{H}$ to $\mathbb{D}$. To start, define a *hyperbolic line in* $\mathbb{D}$ to be the image under m^{-1} of a hyperbolic line in $\mathbb{H}$.

We know that every hyperbolic line in $\mathbb{H}$ is contained in a circle in $\overline{\mathbb{C}}$ perpendicular to $\overline{\mathbb{R}}$, that every element of Möb takes circles in $\overline{\mathbb{C}}$ to circles in $\overline{\mathbb{C}}$, and that every element of Möb preserves the angle between circles in $\overline{\mathbb{C}}$. Hence, every hyperbolic line in $\mathbb{D}$ is the intersection of $\mathbb{D}$ with a circle in $\overline{\mathbb{C}}$ perpendicular to the circle $\mathbb{S}^1$ bounding $\mathbb{D}$.

A picture of some hyperbolic lines in $\mathbb{D}$ is given in Fig. 4.1. Note that this picture of the Poincaré disc model of the hyperbolic plane is vaguely reminiscent of some of the drawings of M. C. Escher. The interested reader is directed to the books of Schattschneider [23] and Locher [17] for more information about the work of Escher.

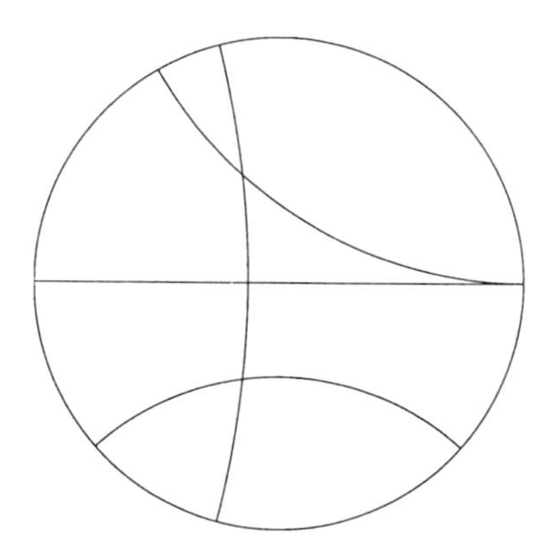

Figure 4.1: Some hyperbolic lines in $\mathbb{D}$

Since the element m of Möb takes $\mathbb{D}$ to $\mathbb{H}$, every element q of Möb($\mathbb{D}$) has the form $q = m^{-1} \circ p \circ m$, where p is an element of Möb($\mathbb{H}$). In particular, the action of Möb($\mathbb{D}$) on $\mathbb{D}$ inherits all the transitivity properties that Möb($\mathbb{H}$) has for its action of $\mathbb{H}$.

In fact, in Exercise 2.38 we saw that every element of $\text{Möb}(\mathbb{D})$ has either the form

$$p(z) = \frac{\alpha z + \beta}{\overline{\beta} z + \overline{\alpha}}$$

or the form

$$p(z) = \frac{\alpha \overline{z} + \beta}{\overline{\beta} \overline{z} + \overline{\alpha}},$$

where α, $\beta \in \mathbb{C}$ and $|\alpha|^2 - |\beta|^2 = 1$. The Möbius transformations taking $\mathbb{D}$ to $\mathbb{D}$ are the elements of

$$\text{Möb}^+(\mathbb{D}) = \text{Möb}^+ \cap \text{Möb}(\mathbb{D}),$$

which are those elements of $\text{Möb}(\mathbb{D})$ of the form

$$p(z) = \frac{\alpha z + \beta}{\overline{\beta} z + \overline{\alpha}}.$$

In order to transfer the hyperbolic element of arc-length from $\mathbb{H}$ to $\mathbb{D}$, we need to have an explicit element n of Möb taking $\mathbb{D}$ to $\mathbb{H}$. The element we use here is

$$n(z) = \frac{\frac{i}{\sqrt{2}} z + \frac{1}{\sqrt{2}}}{-\frac{1}{\sqrt{2}} z - \frac{i}{\sqrt{2}}}.$$

We transfer the hyperbolic element of arc-length from $\mathbb{H}$ to $\mathbb{D}$ by making the following observation. For any piecewise differentiable path $f : [a, b] \to \mathbb{D}$, the composition $n \circ f : [a, b] \to \mathbb{H}$ is a piecewise differentiable path into $\mathbb{H}$. We know how to calculate the hyperbolic length of $n \circ f$, namely by integrating the hyperbolic element of arc-length $\frac{1}{\text{Im}(z)} |dz|$ on $\mathbb{H}$ along $n \circ f$. So, define the hyperbolic length of f in $\mathbb{D}$ by

$$\text{length}_{\mathbb{D}}(f) = \text{length}_{\mathbb{H}}(n \circ f).$$

Theorem 4.1

The hyperbolic length of a piecewise differentiable path $f : [a, b] \to \mathbb{D}$ is given by the integral

$$\text{length}_{\mathbb{D}}(f) = \int_f \frac{2}{1 - |z|^2} \, |dz|.$$

The group of isometries of the resulting hyperbolic metric on $\mathbb{D}$ is $\text{Möb}(\mathbb{D})$.

The proof of Theorem 4.1 consists of several parts. We begin by deriving the form of the hyperbolic element of arc-length on $\mathbb{D}$, and by showing that this hyperbolic element of arc-length is independent of the choice of the element of Möb taking $\mathbb{D}$ to $\mathbb{H}$.

We are given that the hyperbolic length of a piecewise differentiable path $f : [a,b] \to \mathbb{D}$ is given by

$$\begin{aligned}\text{length}_{\mathbb{D}}(f) = \text{length}_{\mathbb{H}}(n \circ f) &= \int_{n\circ f} \frac{1}{\text{Im}(z)} |\mathrm{d}z| \\ &= \int_a^b \frac{1}{\text{Im}((n\circ f)(t))} \, |(n\circ f)'(t)| \, \mathrm{d}t \\ &= \int_a^b \frac{1}{\text{Im}(n(f(t)))} \, |n'(f(t))| \, |f'(t)| \, \mathrm{d}t \\ &= \int_f \frac{1}{\text{Im}(n(z))} |n'(z)| |\mathrm{d}z|.\end{aligned}$$

Calculating, we see that

$$\text{Im}(n(z)) = \text{Im}\left(\frac{\frac{i}{\sqrt{2}}z + \frac{1}{\sqrt{2}}}{-\frac{1}{\sqrt{2}}z - \frac{i}{\sqrt{2}}} \right) = \frac{1 - |z|^2}{|-z-i|^2}$$

and

$$|n'(z)| = \frac{2}{|z+i|^2},$$

and so

$$\frac{1}{\text{Im}(n(z))} |n'(z)| = \frac{2}{1-|z|^2}.$$

We now need to show that this hyperbolic element of arc-length $\frac{2}{1-|z|^2}|\mathrm{d}z|$ on $\mathbb{D}$ is independent of the choice of n. So, let $f : [a,b] \to \mathbb{D}$ be a piecewise differentiable path and let p be any element of Möb taking $\mathbb{D}$ to $\mathbb{H}$.

Since $p \circ n^{-1}$ takes $\mathbb{H}$ to $\mathbb{H}$, we can set $q = p \circ n^{-1}$, so that q is an element of Möb($\mathbb{H}$).

Since $n \circ f$ is a piecewise differentiable path in $\mathbb{H}$, the invariance of hyperbolic length calculated with respect to the element of arc-length $\frac{1}{\text{Im}(z)}|\mathrm{d}z|$ on $\mathbb{H}$ under Möb($\mathbb{H}$) immediately implies that

$$\text{length}_{\mathbb{H}}(n \circ f) = \text{length}_{\mathbb{H}}(q \circ n \circ f) = \text{length}_{\mathbb{H}}(p \circ f).$$

This last equality follows from $q \circ n = p \circ n^{-1} \circ n = p$. Hence, $\text{length}_{\mathbb{D}}(f)$ is well-defined.

As an example calculation, let $0 < r < 1$ and consider the path $f : [0, r] \to \mathbb{D}$ given by $f(t) = t$. Then,

$$\begin{aligned}\text{length}_{\mathbb{D}}(f) &= \int_f \frac{2}{1-|z|^2} \, |\mathrm{d}z| \\ &= \int_0^r \frac{2}{1-t^2} \, \mathrm{d}t\end{aligned}$$

$$
\begin{aligned}
&= \int_0^r \left[\frac{1}{1+t} + \frac{1}{1-t} \right] \mathrm{d}t \\
&= \ln \left[\frac{1+r}{1-r} \right].
\end{aligned}
$$

Exercise 4.1

Let m be an element of Möb taking $\mathbb{H}$ to $\mathbb{D}$ and let $f : [a,b] \to \mathbb{H}$ be a piecewise differentiable path. Show that $\text{length}_{\mathbb{D}}(m \circ f) = \text{length}_{\mathbb{H}}(f)$.

We now use hyperbolic lengths of paths in $\mathbb{D}$ to define hyperbolic distance in $\mathbb{D}$. Given points x and y in $\mathbb{D}$, let $\Theta[x,y]$ be the set of all piecewise differentiable paths $f : [a,b] \to \mathbb{D}$ with $f(a) = x$ and $f(b) = y$, and define

$$\mathrm{d}_{\mathbb{D}}(x,y) = \inf\{\text{length}_{\mathbb{D}}(f) \mid f \in \Theta[x,y]\}.$$

Proposition 4.2

$(\mathbb{D}, \mathrm{d}_{\mathbb{D}})$ is a path metric space. Moreover, a distance realizing path between two points x and y of $\mathbb{D}$ is a parametrization of the hyperbolic line segment joining x to y.

Let m be any element of Möb taking $\mathbb{H}$ to $\mathbb{D}$. The first step of the proof of Proposition 4.2 is to show that m is distance preserving.

As in Section 3.4, let $\Gamma[z,w]$ be the set of all piecewise differentiable paths $f : [a,b] \to \mathbb{H}$ with $f(a) = z$ and $f(b) = w$. For each pair of points z and w of $\mathbb{H}$, we have that

$$
\begin{aligned}
\mathrm{d}_{\mathbb{H}}(z,w) &= \inf\{\text{length}_{\mathbb{H}}(f) \mid f \in \Gamma[z,w]\} \\
&= \inf\{\text{length}_{\mathbb{D}}(m \circ f) \mid f \in \Gamma[z,w]\} \\
&\le \inf\{\text{length}_{\mathbb{D}}(g) \mid g \in \Theta[m(z), m(w)]\} \\
&\le \mathrm{d}_{\mathbb{D}}(m(z), m(w)).
\end{aligned}
$$

Similarly, if x and y are points of $\mathbb{D}$, write $x = m(z)$ and $y = m(w)$ for points z and w of $\mathbb{H}$. Calculating, we see that

$$
\begin{aligned}
\mathrm{d}_{\mathbb{D}}(x,y) = \mathrm{d}_{\mathbb{D}}(m(z), m(w)) &= \inf\{\text{length}_{\mathbb{D}}(f) \mid f \in \Theta[x,y]\} \\
&= \inf\{\text{length}_{\mathbb{H}}(m \circ f) \mid f \in \Theta[x,y]\} \\
&\le \inf\{\text{length}_{\mathbb{H}}(g) \mid g \in \Theta[z,w]\} \\
&\le \mathrm{d}_{\mathbb{H}}(z,w).
\end{aligned}
$$

Since m is a distance-preserving homeomorphism between $\mathbb{H}$ and $\mathbb{D}$, and since $\mathrm{d}_{\mathbb{H}}$ is a metric on $\mathbb{H}$, we have that $\mathrm{d}_{\mathbb{D}}$ is a metric on $\mathbb{D}$.

To complete the proof of Proposition 4.2, let x and y be two points of $\mathbb{D}$, let $z = m^{-1}(x)$ and $w = m^{-1}(y)$, and let $f : [a, b] \to \mathbb{H}$ be a piecewise differentiable path with $f(a) = z$, $f(b) = w$, and $\mathrm{length}_{\mathbb{H}}(f) = \mathrm{d}_{\mathbb{H}}(z, w)$.

Since m is a distance preserving homeomorphism between $\mathbb{H}$ and $\mathbb{D}$, there necessarily exists a path in $\Theta[x, y]$ realizing the hyperbolic distance $\mathrm{d}_{\mathbb{D}}(x, y)$, namely $m \circ f$.

Moreover, since f is a parametrization of the hyperbolic line segment in $\mathbb{H}$ between z and w, and since m takes hyperbolic lines in $\mathbb{H}$ to hyperbolic lines in $\mathbb{D}$, we see that $m \circ f$ is the parametrization of the hyperbolic line segment in $\mathbb{D}$ between x and y. Hence, in $\mathbb{D}$ as in $\mathbb{H}$, the distance realizing path between two points is a parametrization of the hyperbolic line segment joining them. This completes the proof of Proposition 4.2.

Exercise 4.2

For $0 < r < 1$, show that

$$\mathrm{d}_{\mathbb{D}}(0, r) = \ln \left[\frac{1 + r}{1 - r} \right],$$

and hence that

$$r = \tanh \left[\frac{1}{2} \mathrm{d}_{\mathbb{D}}(0, r) \right].$$

The fact that $\mathrm{M\ddot{o}b}(\mathbb{D})$ is exactly the group of isometries of $(\mathbb{D}, \mathrm{d}_{\mathbb{D}})$ follows from the fact that $\mathrm{M\ddot{o}b}(\mathbb{H})$ is exactly the group of isometries of $(\mathbb{H}, \mathrm{d}_{\mathbb{H}})$, by Theorem 3.12, and that any element m of Möb taking $\mathbb{H}$ to $\mathbb{D}$ is a distance-preserving homeomorphism, and hence an isometry.

Specifically, if g is an isometry of $(\mathbb{D}, \mathrm{d}_{\mathbb{D}})$, then $m^{-1} \circ g \circ m$ is an isometry of $(\mathbb{H}, \mathrm{d}_{\mathbb{H}})$. By Theorem 3.12, we have that $m^{-1} \circ g \circ m$ is an element of $\mathrm{M\ddot{o}b}(\mathbb{H})$, and hence g is an element of $\mathrm{M\ddot{o}b}(\mathbb{D})$.

Conversely, if g is an element of $\mathrm{M\ddot{o}b}(\mathbb{D})$, then $m^{-1} \circ g \circ m$ is an element of $\mathrm{M\ddot{o}b}(\mathbb{H})$, and hence is an isometry of $(\mathbb{H}, \mathrm{d}_{\mathbb{H}})$. Since m and m^{-1} are distance-preserving, we have that g is an isometry of $(\mathbb{D}, \mathrm{d}_{\mathbb{D}})$. This completes the proof of Theorem 4.1.

We note here that, analogously to the upper half-plane $\mathbb{H}$, the *boundary at infinity* of the Poincaré disc $\mathbb{D}$ is the unit circle $\mathbb{S}^1$ in $\mathbb{C}$, which is the circle in $\overline{\mathbb{C}}$ determining $\mathbb{D}$. As with the boundary at infinity $\overline{\mathbb{R}}$ of $\mathbb{H}$, the hyperbolic distance between a point of $\mathbb{S}^1$ and a point of $\mathbb{D}$ is infinite.

One difficulty with the upper half-plane model $\mathbb{H}$ of the hyperbolic plane is that there is no easily expressed relationship between the Euclidean distance $|z-w|$ and the hyperbolic distance $\mathrm{d}_{\mathbb{H}}(z,w)$.

One of the useful features of the Poincaré disc model $\mathbb{D}$ is that there does exist an easily expressed relationship between the Euclidean and hyperbolic distance between a pair of points of $\mathbb{D}$. We find this relationship by considering functions on $\mathbb{D}$ which are invariant under $\text{Möb}^+(\mathbb{D})$.

This is very similar to the discussion in Section 2.3. Say that a function g from $\mathbb{D}\times\mathbb{D}$ to $\mathbb{R}$ is *invariant under the action of* $\text{Möb}^+(\mathbb{D})$ if for each point (x,y) of $\mathbb{D}\times\mathbb{D}$ and for each element p of $\text{Möb}^+(\mathbb{D})$, we have that $g(x,y)=g(p(x),p(y))$.

We already know one such function, namely the hyperbolic distance $\mathrm{d}_{\mathbb{D}}$. In fact, $\mathrm{d}_{\mathbb{D}}$ is invariant under the larger group $\text{Möb}(\mathbb{D})$. It is easy to see that for any function $h:[0,\infty)\to\mathbb{R}$, the composition $\varphi=h\circ\mathrm{d}_{\mathbb{D}}$ is invariant under $\text{Möb}(\mathbb{D})$. Let us try and find an explicit example.

To begin with, the invariance of hyperbolic lengths of paths in $\mathbb{D}$ under the action of $\text{Möb}^+(\mathbb{D})$ gives that

$$\begin{aligned}\int_a^b \frac{2}{1-|f(t)|^2}|f'(t)|\mathrm{d}t &= \int_a^b \frac{2}{1-|(p\circ f)(t)|^2}|(p\circ f)'(t)|\mathrm{d}t\\ &= \int_a^b \frac{2}{1-|p(f(t))|^2}\,|p'(f(t))|\,|f'(t)|\mathrm{d}t\end{aligned}$$

for every piecewise differentiable path $f:[a,b]\to\mathbb{D}$ and every element p of $\text{Möb}^+(\mathbb{D})$.

Since this holds for every piecewise differentiable path $f:[a,b]\to\mathbb{D}$, we may use Lemma 3.4 to conclude that

$$\frac{2}{1-|z|^2}=\frac{2|p'(z)|}{1-|p(z)|^2}$$

for every element p of $\text{Möb}^+(\mathbb{D})$.

We now calculate that

$$(p(x)-p(y))^2=p'(x)p'(y)(x-y)^2$$

for every element p of $\text{Möb}^+(\mathbb{D})$ and every pair x and y of points of $\mathbb{D}$.

Namely, we write

$$p(z)=\frac{\alpha z+\beta}{\overline{\beta}z+\overline{\alpha}},$$

where $\alpha, \beta \in \mathbb{C}$ and $|\alpha|^2 - |\beta|^2 = 1$. Then,

$$p(z) - p(w) = \frac{z - w}{(\overline{\beta}z + \overline{\alpha})(\overline{\beta}w + \overline{\alpha})}$$

and

$$p'(z) = \frac{1}{(\overline{\beta}z + \overline{\alpha})^2}.$$

Combining these two calculations, we can see that

$$\begin{aligned} \frac{|x-y|^2}{(1-|x|^2)(1-|y|^2)} &= |x-y|^2 \left(\frac{|p'(x)|}{1-|p(x)|^2}\right)\left(\frac{|p'(y)|}{1-|p(y)|^2}\right) \\ &= \frac{|p(x)-p(y)|^2}{(1-|p(x)|^2)(1-|p(y)|^2)}. \end{aligned}$$

Consequently, the function $\varphi : \mathbb{D} \times \mathbb{D} \to \mathbb{R}$ defined by

$$\varphi(x, y) = \frac{|x-y|^2}{(1-|x|^2)(1-|y|^2)}$$

is invariant under the action of $\text{Möb}^+(\mathbb{D})$.

The main application of the invariance of φ under the action $\text{Möb}^+(\mathbb{D})$ is to provide a link between the Euclidean and hyperbolic distances between a pair of points of $\mathbb{D}$.

Proposition 4.3

For each pair x and y of points of $\mathbb{D}$, we have that

$$\varphi(x, y) = \sinh^2\left(\frac{1}{2}\mathrm{d}_{\mathbb{D}}(x, y)\right) = \frac{1}{2}\left(\cosh(\mathrm{d}_{\mathbb{D}}(x, y)) - 1\right).$$

The proof of Proposition 4.3 is by direct calculation. Let x and y be a pair of points in $\mathbb{D}$. Choose an element $p(z) = \frac{\alpha z+\beta}{\overline{\beta}z+\overline{\alpha}}$ of $\text{Möb}^+(\mathbb{D})$ (so that $\alpha, \beta \in \mathbb{C}$ and $|\alpha|^2 - |\beta|^2 = 1$) for which $p(x) = 0$.

One way to do this is to set $\beta = -\alpha x$, so that

$$p(z) = \frac{\alpha(z - x)}{\overline{\alpha}(-\overline{x}z + 1)},$$

where $|\alpha|^2(1 - |x|^2) = 1$. Now choose the argument of α so that $p(y) = r$ is real and positive. Then,

$$\begin{aligned} \frac{|x-y|^2}{(1-|x|^2)(1-|y|^2)} &= \varphi(x, y) \\ &= \varphi(p(x), p(y)) = \varphi(0, r) = \frac{r^2}{1-r^2}. \end{aligned}$$

Since by Exercise 4.2 we have that $r = \tanh(\frac{1}{2}\mathrm{d}_{\mathbb{D}}(0,r))$, we see that

$$\frac{r^2}{1-r^2} = \sinh^2\left(\frac{1}{2}\mathrm{d}_{\mathbb{D}}(x,y)\right) = \frac{1}{2}\left(\cosh(\mathrm{d}_{\mathbb{D}}(x,y)) - 1\right),$$

as desired. This completes the proof of Proposition 4.3.

Exercise 4.3

Let ℓ_1 and ℓ_2 be two intersecting hyperbolic lines in $\mathbb{D}$, where the endpoints at infinity of ℓ_1 are z_1 and z_2, and the endpoints at infinity of ℓ_2 are w_1 and w_2, labelled so that the order of the points counter-clockwise around $\mathbb{S}^1$ is z_1, w_1, z_2, w_2. Prove that the angle θ between ℓ_1 and ℓ_2 satisfies

$$[z_1, w_1; z_2, w_2]\tan^2\left(\frac{\theta}{2}\right) = -1.$$

We close this section with a discussion of hyperbolic circles.

Definition 4.4

A *hyperbolic circle* in $\mathbb{D}$ is a set in $\mathbb{D}$ of the form

$$C = \{y \in \mathbb{D} \mid \mathrm{d}_{\mathbb{D}}(x,y) = s\},$$

where $x \in \mathbb{D}$ and $s > 0$ are fixed. We refer to x as the *hyperbolic centre* of C and s as the *hyperbolic radius* of C.

We are able to completely characterize hyperbolic circles in $\mathbb{D}$.

Proposition 4.5

A hyperbolic circle in $\mathbb{D}$ is a Euclidean circle in $\mathbb{D}$ and vice versa, through the hyperbolic and Euclidean centres, and the hyperbolic and Euclidean radii, will in general be different.

We begin with a specific set of hyperbolic circles in $\mathbb{D}$, namely those centred at 0. Given $s > 0$, set $r = \tanh(\frac{1}{2}s)$, so that $\mathrm{d}_{\mathbb{D}}(0,r) = s$.

Since Möb($\mathbb{D}$) contains $e(z) = e^{i\theta}z$, we see that e is an isometry of $(\mathbb{D}, \mathrm{d}_{\mathbb{D}})$, and so every point $re^{i\theta}$ in $\mathbb{D}$ satisfies $\mathrm{d}_{\mathbb{D}}(0, re^{i\theta}) = s$ as well.

Hence, the Euclidean circle with Euclidean centre 0 and Euclidean radius r and the hyperbolic circle with hyperbolic centre 0 and hyperbolic radius s are the same, where s and r are related by $r = \tanh(\frac{1}{2}s)$.

Let C be the hyperbolic circle in $\mathbb{D}$ with hyperbolic centre c and hyperbolic radius s. Let m be an element of Möb($\mathbb{D}$) taking c to 0. Then, $m(C)$ is the hyperbolic circle in $\mathbb{D}$ with hyperbolic centre 0 and hyperbolic radius s. In particular, $m(C)$ is also a Euclidean circle.

Since the elements of Möb($\mathbb{D}$) take circles in $\overline{\mathbb{C}}$ to circles in $\overline{\mathbb{C}}$, we see that C is also a circle in $\overline{\mathbb{C}}$. Since no element of Möb($\mathbb{D}$) takes a point of $\mathbb{D}$ to ∞, we see that C is necessarily a Euclidean circle in $\mathbb{D}$.

Similarly, if C is a Euclidean circle in $\mathbb{D}$, we choose an element of Möb($\mathbb{D}$) taking C to a Euclidean circle in $\mathbb{D}$ with Euclidean centre 0, and so C is also a hyperbolic circle in $\mathbb{D}$.

Exercise 4.4

Given $s > 0$, let S_s be the hyperbolic circle in $\mathbb{D}$ with hyperbolic centre 0 and hyperbolic radius s. Show that the hyperbolic length of S_s is

$$\text{length}_{\mathbb{D}}(S_s) = 2\pi \sinh(s).$$

4.2 A General Construction

The construction from Section 4.1, of transferring hyperbolic geometry from the upper half-plane $\mathbb{H}$ to the unit disc $\mathbb{D}$, is actually just a single instance of a more general method of constructing models of the hyperbolic plane from the upper half-plane model $\mathbb{H}$.

We work in a somewhat restricted setting. Let X be a subset of $\mathbb{C}$ that is *diffeomorphic* to $\mathbb{H}$. That is, let X be any subset of $\mathbb{C}$ for which there exists a homeomorphism $\xi : X \to \mathbb{H}$ so that ξ and its inverse ξ^{-1} are both differentiable as functions of x and y, as described in Note 3.3.

One example of this sort of function that we have already seen, in Section 4.1, is to take $X = \mathbb{D}$ and to write

$$\xi(z) = \frac{iz+1}{-z-i} = \frac{-2x}{x^2+(y+1)^2} + i\frac{1-x^2-y^2}{x^2+(y+1)^2}.$$

In a very crude fashion, we may use ξ to transfer the hyperbolic geometry from $\mathbb{H}$ to X and so to get a model of the hyperbolic plane whose underlying space is X. Specifically, define a *hyperbolic line in* X to be the image in X of a hyperbolic line in $\mathbb{H}$ under ξ^{-1}.

While having a description of the hyperbolic lines in X is nice, it is not in general easy to work with. Of more use would be to use ξ to transfer the hyperbolic element of arc-length $\frac{1}{\mathrm{Im}(z)}|dz|$ on $\mathbb{H}$ to a hyperbolic element of arc-length on X, so that we may actually calculate in this new model of the hyperbolic plane with underlying space X.

We accomplish this transfer of the hyperbolic element of arc-length from $\mathbb{H}$ to X exactly as we transferred the hyperbolic element of arc-length from $\mathbb{H}$ to $\mathbb{D}$.

That is, define the hyperbolic element of arc-length ds_X on X by declaring that

$$\mathrm{length}_X(f) = \int_f ds_X = \int_{\xi\circ f} \frac{1}{\mathrm{Im}(z)}\,|dz| = \mathrm{length}_{\mathbb{H}}(\xi\circ f)$$

for every piecewise differentiable path $f : [a,b] \to X$.

We note that, in general, there may not be a nice form for ds_X. For example, let $X = \mathbb{C}$ and consider the diffeomorphism $\xi : \mathbb{C} \to \mathbb{H}$ given by

$$\xi(z) = \mathrm{Re}(z) + \exp(\mathrm{Im}(z))i.$$

Let $f : [a,b] \to \mathbb{C}$ be a piecewise differentiable path, and write $f(t) = x(t) + iy(t)$. Calculating, we see that

$$\xi\circ f(t) = x(t) + \exp(y(t))i.$$

So,

$$\mathrm{Im}((\xi\circ f)(t)) = \exp(y(t))$$

and

$$|(\xi\circ f)'(t)| = \sqrt{(x'(t))^2 + (y'(t))^2\exp(2y(t))},$$

and so

$$\mathrm{length}_X(f) = \int_f ds_X = \int_a^b \frac{1}{\exp(y(t))}\sqrt{(x'(t))^2 + (y'(t))^2\exp(2y(t))}\,dt.$$

Even if we consider the particular path $f : [0,2\pi] \to \mathbb{C}$ given by $f(t) = s\exp(it)$, so that $x(t) = s\cos(t)$ and $y(t) = s\sin(t)$, we get that

$$\mathrm{length}_X(f) = \int_a^b \frac{1}{\exp(s\sin(t))}\sqrt{s^2\sin^2(t) + s^2\cos^2(t)\exp(2s\sin(t))}\,dt.$$

So, in general, this does not seem to be an effective general method for constructing models of the hyperbolic plane, as the element of arc-length is not easily expressible. This construction of the element of arc-length $\mathrm{d}s_X$ on X using the diffeomorphism $\xi : X \to \mathbb{H}$ is often referred to as defining $\mathrm{d}s_X$ to be the *pullback* of the element of arc-length on $\mathbb{H}$ by ξ.

Exercise 4.5

Consider the diffeomorphism $\xi : \mathbb{H} \to \mathbb{H}$ given by

$$\xi(z) = \mathrm{Re}(z) + \frac{1}{2}\mathrm{Im}(z)i.$$

Calculate the pullback of $\frac{1}{\mathrm{Im}(z)}|\mathrm{d}z|$ by ξ.

There is a special case in which it is significantly easier to write down the element of arc-length $\mathrm{d}s_X$ on X, which includes the case of the Poincaré disc $\mathbb{D}$. Suppose that X is an open subset of the complex plane $\mathbb{C}$ and that $\xi : X \to \mathbb{H}$ is a diffeomorphism that is differentiable as a function of z, as described in Note 3.3.

Theorem 4.6

Suppose that X is an open subset of the complex plane $\mathbb{C}$ and that $\xi : X \to \mathbb{H}$ is a diffeomorphism that is differentiable as a function of z. The pullback $\mathrm{d}s_X$ of the hyperbolic element of arc-length $\frac{1}{\mathrm{Im}(z)}\,|\mathrm{d}z|$ on $\mathbb{H}$ is

$$\mathrm{d}s_X = \frac{1}{\mathrm{Im}(\xi(z))}|\xi'(z)||\mathrm{d}z|.$$

The proof of Theorem 4.6 is a direct calculation. Proceeding as above, let $f : [a, b] \to X$ be a piecewise differentiable path. The hyperbolic length of f is given by

$$\begin{aligned}
\mathrm{length}_X(f) = \int_f \mathrm{d}s_X &= \int_{\xi \circ f} \frac{1}{\mathrm{Im}(z)}|\mathrm{d}z| \\
&= \int_a^b \frac{1}{\mathrm{Im}(\xi(f(t)))}|\xi'(f(t))||f'(t)|\mathrm{d}t \\
&= \int_f \frac{1}{\mathrm{Im}(\xi(z))}|\xi'(z)||\mathrm{d}z|.
\end{aligned}$$

Applying Lemma 3.4 completes the proof of Theorem 4.6.

As an example, let

$$X = \{z \in \mathbb{C} \mid \mathrm{Re}(z) > 0 \text{ and } \mathrm{Im}(z) > 0\},$$

and consider the diffeomorphism $\xi : X \to \mathbb{H}$ given by $\xi(z) = z^2$.

Since

$$\mathrm{Im}(\xi(z)) = \mathrm{Im}(z^2) = 2\,\mathrm{Re}(z)\,\mathrm{Im}(z)$$

and

$$|\xi'(z)| = |2z| = 2|z|,$$

we see that the pullback of $\frac{1}{\mathrm{Im}(z)}|\mathrm{d}z|$ by ξ is

$$\mathrm{d}s_X = \frac{1}{\mathrm{Im}(\xi(z))}|\xi'(z)||\mathrm{d}z| = \frac{|z|}{\mathrm{Re}(z)\,\mathrm{Im}(z)}|\mathrm{d}z|.$$

Moreover, we can explicitly describe the hyperbolic lines in this model X. If we let $w = u + iv$ be the coordinate on X, then $\xi(w) = u^2 - v^2 + 2iuv$. The hyperbolic lines in $\mathbb{H}$ are of two types, those contained in the Euclidean line $L_c = \{z \in \mathbb{H} \mid \mathrm{Re}(z) = c\}$ and those contained in the Euclidean circle $A_{c,r} = \{z \in \mathbb{H} \mid (\mathrm{Re}(z) - c)^2 + (\mathrm{Im}(z))^2 = r^2\}$.

The image of L_c under ξ^{-1} is the curve $\{w \in X \mid u^2 - v^2 = c\}$ in X. For $c = 0$, this curve is the Euclidean ray K from 0 making angle $\frac{\pi}{4}$ with the positive real axis, while for $c \neq 0$ this curve is a hyperboloid asymptotic to K.

The image of $A_{c,r}$ under ξ^{-1} is a curve known as an *oval of Cassini*, given by the equation

$$(u^2 + v^2)^2 - 2c(u^2 - v^2) + c^2 = r^2.$$

An oval of Cassini is a variant on an ellipse. Let w_0 and w_1 be two fixed points in $\mathbb{C}$. While an ellipse is the set of points in $\mathbb{C}$ for which the sum of (Euclidean) distances $|w - w_0| + |w - w_1|$ is constant, an oval of Cassini is the set of points w in $\mathbb{C}$ for which the product of the (Euclidean) distances $|(w - w_0)(w - w_1)|$ is constant. Unlike ellipses, all of which have the same shape, the shape of ovals of Cassini can change as the value of this constant changes.

Exercise 4.6

Let $Y = \{z \in \mathbb{C} | \mathrm{Re}(z) > 0\}$, and consider the diffeomorphism $\xi : Y \to \mathbb{H}$ given by $\xi(z) = i\,z$. Determine the pullback of $\frac{1}{\mathrm{Im}(z)}|\mathrm{d}z|$ by ξ.

In exactly the same way that we defined the hyperbolic metric on $\mathbb{D}$ and determined its group of isometries, this construction allows us to define the hyperbolic metric on X and determine its group of isometries.

Specifically, let X be an open subset of $\mathbb{C}$ and let $\xi : X \to \mathbb{H}$ be a diffeomorphism that is differentiable as a function of z. Let $\mathrm{d}s_X$ be the pullback of the hyperbolic element $\frac{1}{\mathrm{Im}(z)}|\mathrm{d}z|$ on $\mathbb{H}$ by ξ. Then, we can use $\mathrm{d}s_X$ to define a hyperbolic metric d_X on X by taking the infimum of hyperbolic lengths of paths in X.

Using the same proofs as we used in Section 4.1 for the Poincaré disc $\mathbb{D}$, we see that (X, d_X) is a path metric space whose distance realizing paths are precisely the parametrizations of the hyperbolic line segments in X. Also, the group of isometries of (X, d_X) is

$$\mathrm{Isom}(X, d_X) = \{\xi^{-1} \cdot m \cdot \xi \mid m \in \mathrm{M\ddot{o}b}(\mathbb{H})\}.$$

We close this section by introducing the notion of curvature. Let X be an open subset of $\mathbb{C}$ and let $ds_X = \alpha(z)|dz|$ be an element of arc-length on X, where α is a positive real-valued differentiable function on X. By the same constructions we have seen several times, this element of arc-length induces a metric on X, where the distance between two points is the infimum of the lengths of all paths joining the two points, where the length of a path is calculated with respect to this element of arc-length.

There is a numerical quantity associated to the metric that arises from this element of arc-length, called the *curvature* of the metric. The study of metrics and their properties, such as curvature, is properly the subject of Differential Geometry, but we say a few words here.

The curvature of the metric induced by $ds_X = \alpha(z)|dz|$ is itself a function $\mathrm{curv} : X \to \mathbb{R}$, which is given explicitly by the formula

$$\mathrm{curv}(z) = -\left[\frac{2}{\alpha(z)}\right]^2 \partial\overline{\partial}\log(\alpha(z)).$$

Here, if we write $z = x + iy$ and $\beta(z) = f(x, y) + ig(x, y)$, we set

$$\partial\beta = \frac{1}{2}\left[\frac{\partial}{\partial x}\beta - i\frac{\partial}{\partial y}\beta\right] = \frac{1}{2}\left[\frac{\partial f}{\partial x} + \frac{\partial g}{\partial y} + i\left(\frac{\partial g}{\partial x} - \frac{\partial f}{\partial y}\right)\right]$$

and

$$\overline{\partial}\beta = \frac{1}{2}\left[\frac{\partial}{\partial x}\beta + i\frac{\partial}{\partial y}\beta\right] = \frac{1}{2}\left[\frac{\partial f}{\partial x} - \frac{\partial g}{\partial y} + i\left(\frac{\partial g}{\partial x} + \frac{\partial f}{\partial y}\right)\right].$$

Exercise 4.7

Check that

$$\partial\overline{\partial}\beta = \frac{1}{4}\left[\frac{\partial^2\beta}{\partial x^2} + \frac{\partial^2\beta}{\partial y^2}\right].$$

In particular, note that for the standard Euclidean metric on $\mathbb{C}$, we have $\alpha \equiv 1$, and so the curvature of the Euclidean metric on $\mathbb{C}$ is identically zero.

For the upper half-plane model $\mathbb{H}$, the element of arc-length is $\frac{1}{\text{Im}(z)}|dz|$, and so the curvature function is

$$\text{curv}(z) = -4\,(\text{Im}(z))^2\,\partial\overline{\partial}\log\left(\frac{1}{\text{Im}(z)}\right) = -1.$$

Note that if we consider the slightly more general hyperbolic element of arc-length $\frac{c}{\text{Im}(z)}|dz|$ on $\mathbb{H}$, we have that the curvature is $-\frac{1}{c^2}$.

Exercise 4.8

Calculate the curvature of the hyperbolic metric on $\mathbb{D}$ coming from the hyperbolic element of arc-length $\frac{2}{1-|z|^2}|dz|$.

Exercise 4.9

Calculate the curvature of the metric on $\mathbb{C}$ coming from the element of arc-length $\frac{1}{1+|z|^2}|dz|$.

In fact, let X be any *connected* and *simply connected* open subset of $\mathbb{C}$. Roughly, a connected set is a set which has only one piece.

Definition 4.7

An open set X in $\mathbb{C}$ is *connected* if, given any two points x and y in X, there exists a piecewise differentiable path $f : [a,b] \to X$ with $f(a) = x$ and $f(y) = y$.

Roughly, a simply connected set does not have any holes. Before defining simply connected, we need to know what a *Jordan curve* is.

Definition 4.8

A *Jordan curve* C is a simple closed curve in $\mathbb{C}$. That is, there exists a continuous path $f : [0,2\pi] \to \mathbb{C}$ so that $f(0) = f(2\pi)$, f is injective on $[0,2\pi)$, and $f([0,2\pi]) = C$. The unit circle $\mathbb{S}^1$ in $\mathbb{C}$ is an example of a Jordan curve.

The Jordan curve theorem states that complement in $\mathbb{C}$ of a Jordan curve C has exactly two components, one bounded and the other unbounded. We refer to the bounded component of $\mathbb{C} - C$ as the *disc* bounded by C.

Definition 4.9

A connected open set X in $\mathbb{C}$ is *simply connected* if for every Jordan curve C, the disc bounded by C is also contained in X.

Both the upper half-plane $\mathbb{H}$ and the unit disc $\mathbb{D}$ are connected and simply connected. The punctured plane $\mathbb{C} - \{0\}$ is connected but is not simply connected, since the unit circle $\mathbb{S}^1$ is contained in $\mathbb{C} - \{0\}$ but the disc bounded by $\mathbb{S}^1$, namely the unit disc $\mathbb{D}$, is not contained in $\mathbb{C} - \{0\}$.

The classical uniformization theorem yields that there are only two possibilities for a connected and simply connected open subset X of $\mathbb{C}$. One is that $X = \mathbb{C}$. The other is that there exists a diffeomorphism $\xi : X \to \mathbb{H}$ which is differentiable as a function of z, and so it is possible to put a hyperbolic element of arc-length on X, and hence to do hyperbolic geometry on X.

An exact statement of the classical uniformization theorem is beyond the scope of this book. A very good exposition can be found in the article of Abikoff [1] and the sources contained in its bibliography.

5
Convexity, Area, and Trigonometry

In this chapter, we explore some of the finer points of hyperbolic geometry. We first describe the notion of *convexity* and explore *convex sets*, including the class of *hyperbolic polygons*. Restricting our attention to hyperbolic polygons, we go on to discuss the measurement of *hyperbolic area*, including the *Gauss-Bonnet formula*, which gives a formula for the hyperbolic area of a hyperbolic polygon in terms of its angles. We go on to use the Gauss-Bonnet formula to show that non-trivial *dilations* of the hyperbolic plane do not exist. We close the chapter with a discussion of the *laws of trigonometry* in the hyperbolic plane.

5.1 Convexity

We now have a good working knowledge of the geometry of the hyperbolic plane. We have several different models to work in, and we have fairly explicit descriptions of hyperbolic length and hyperbolic distance in these models. We now begin to explore some of the finer points of hyperbolic geometry.

In this section, we consider the notion of *convexity* . Recall that we know what it means for a set Z in the complex plane $\mathbb{C}$ to be convex, namely that for each pair z_0 and z_1 of points in Z, the Euclidean line segment joining z_0 and z_1 also lies in Z. In $\mathbb{C}$, this can be expressed formulaically by saying that Z is convex if for each pair z_0 and z_1 of points of Z, all of the points $z_t = (1-t)z_0 + tz_1$ for $0 \leq t \leq 1$ also lie in Z.

We can consider this definition in the hyperbolic plane.

Definition 5.1

A subset X of the hyperbolic plane is *convex* if for each pair of points x and y in X, the closed hyperbolic line segment ℓ_{xy} joining x to y is contained in X.

Unlike in the complex plane, in the models of the hyperbolic plane we have encountered so far, there is not in general a nice parametrization of the hyperbolic line segment joining two arbitrary points.

Note that since convexity is defined in terms of hyperbolic line segments, it is an immediate consequence of the definition that convexity is preserved by hyperbolic isometries. That is, if X is a convex set in the hyperbolic plane and if γ is an isometry of the hyperbolic plane, then $\gamma(X)$ is also convex.

An easy example of a convex set in the hyperbolic plane is a hyperbolic line.

Proposition 5.2

Hyperbolic lines, hyperbolic rays, and hyperbolic line segments are convex.

Let ℓ be a hyperbolic line, and let x and y be two points of ℓ. By Proposition 1.2, x and y determine a unique hyperbolic line, namely ℓ, and so the closed hyperbolic line segment ℓ_{xy} joining x to y is necessarily contained in ℓ. Hence, ℓ is convex.

This same argument also shows that hyperbolic rays and hyperbolic line segments are convex. This completes the proof of Proposition 5.2.

Convexity behaves well under intersections.

Exercise 5.1

Suppose that $\{X_\alpha\}_{\alpha \in A}$ is a collection of convex subsets of the hyperbolic plane. Prove that the intersection $X = \cap_{\alpha \in A} X_\alpha$ is convex.

Another example, and in a sense the most basic example, of a convex set in the hyperbolic plane is a *half-plane*, as discussed in Section 1.2. To recall the definition, given a hyperbolic line ℓ in the hyperbolic plane, the complement of ℓ in the hyperbolic plane has two components, which are the two *open half-planes determined by* ℓ.

A *closed half-plane determined by* ℓ is the union of ℓ with one of the two open half-planes determined by ℓ. We often refer to ℓ as the *bounding line* for the half-planes it determines. We describe why half-planes can be thought of as the most basic convex sets in Section 5.2.

We now show that half-planes are convex.

Proposition 5.3

Open half-planes and closed half-planes in the hyperbolic plane are convex.

We work in the upper half-plane model $\mathbb{H}$, and we begin with a specific half-plane. Let I be the positive imaginary axis in $\mathbb{H}$, and consider the open half-plane

$$U = \{z \in \mathbb{H} \mid \text{Re}(z) > 0\}$$

in $\mathbb{H}$ determined by I.

Let x and y be two points of U. If $\text{Re}(x) = \text{Re}(y)$, then the hyperbolic line segment ℓ_{xy} joining x to y is contained in the Euclidean line $L = \{z \in \mathbb{H} | \text{Re}(z) = \text{Re}(x)\}$. Since L is parallel to I, we see that ℓ_{xy} is disjoint from I, and so ℓ_{xy} is contained in U.

If $\text{Re}(x) \neq \text{Re}(y)$, then the hyperbolic line segment ℓ_{xy} joining x to y lies in the Euclidean circle C with centre on the real axis $\mathbb{R}$. Since the intersection of C and I contains at most one point, and since both x and y are contained in U, we have that ℓ_{xy} is contained in U.

So, U is convex. Combining this argument with the fact that $\text{Möb}(\mathbb{H})$ acts transitively on the set of open half-planes of $\mathbb{H}$, by Exercise 2.41, and the fact that $\text{Möb}(\mathbb{H})$ preserves convexity, we have that every open half-plane in $\mathbb{H}$ is convex.

We may repeat this argument without change with a closed half-plane, and obtain that closed half-planes are convex as well. This completes the proof of Proposition 5.3.

Exercise 5.2

Prove that the open hyperbolic disc D_s in the Poincaré disc $\mathbb{D}$ with hyperbolic centre 0 and hyperbolic radius $s > 0$ is convex. Conclude that all hyperbolic discs are convex.

On the other hand, convexity does not behave well under unions. To take one example, let ℓ_1 and ℓ_2 be two distinct, though not necessarily disjoint, hyperbolic lines.

Take points z_1 on ℓ_1 and z_2 on ℓ_2, chosen only so that neither z_1 nor z_2 is the point of intersection $\ell_1 \cap \ell_2$ of ℓ_1 and ℓ_2. Then, the hyperbolic line segment ℓ_{12} joining z_1 to z_2 does not lie in $\ell_1 \cup \ell_2$, and so $\ell_1 \cup \ell_2$ is not convex.

For an illustration of this phenomenon for the two hyperbolic lines in $\mathbb{H}$ contained in the positive imaginary axis and the unit circle, see Fig. 5.1.

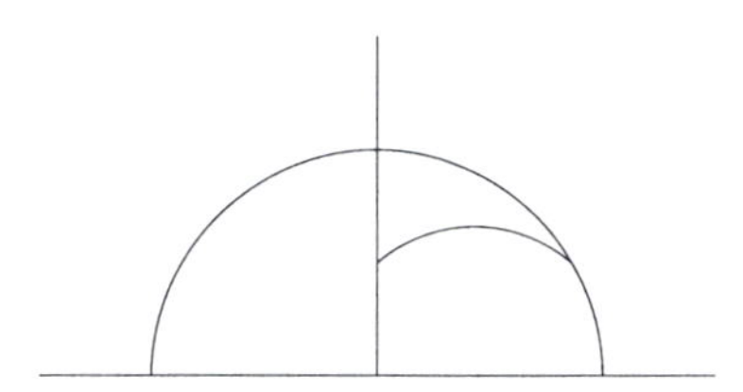

Figure 5.1: The non-convex union of two hyperbolic lines

Adding the hypothesis of convexity allows us to refine some of the results in Section 3.7 about the properties of the hyperbolic metric and the hyperbolic distance between sets. For example, consider the following generalization of Exercise 3.18.

Proposition 5.4

Let X be a closed, convex subset of $\mathbb{H}$, and let z be a point of $\mathbb{H}$ not in X. Then, there exists a unique point $x \in X$ with $d_{\mathbb{H}}(z, x) = d_{\mathbb{H}}(z, X)$.

We first show that there exists some point x of X with $d_{\mathbb{H}}(z, x) = d_{\mathbb{H}}(z, X)$. Since $d_{\mathbb{H}}(z, X) = \inf\{d_{\mathbb{H}}(z, x) \mid x \in X\}$, there exists a sequence $\{x_n\}$ of points of X so that

$$\lim_{n\to\infty} d_{\mathbb{H}}(z, x_n) = d_{\mathbb{H}}(z, X).$$

In particular, by the definition of convergence there is some $N > 0$ so that $d_{\mathbb{H}}(z, x_n) \leq d_{\mathbb{H}}(z, X) + 1$ for $n \geq N$. Set $C = d_{\mathbb{H}}(z, X) + 1$, and let

$$V_C(z) = \{w \in \mathbb{H} \mid d_{\mathbb{H}}(z, w) \leq C\}$$

denote the closed hyperbolic disc with hyperbolic centre z and hyperbolic radius C.

The subset $X \cap V_C(z)$ of $\mathbb{H}$ is closed and bounded, and hence is compact. Since $\{x_n \mid n \geq N\}$ is a sequence contained in the compact subset $X \cap V_C(z)$ of $\mathbb{H}$, there exists a subsequence $\{x_{n_k}\}$ of $\{x_n \mid n \geq N\}$ that converges to some point x of $\mathbb{H}$. Since each x_{n_k} is contained in X and since X is closed, we have that $x \in X$.

Since

$$\mathrm{d}_{\mathbb{H}}(z, x) = \lim_{k \to \infty} \mathrm{d}_{\mathbb{H}}(z, x_{n_k}),$$

we have that $\mathrm{d}_{\mathbb{H}}(z, x) = \mathrm{d}_{\mathbb{H}}(z, X)$.

We need now to show that this point x is unique. So, suppose there are two points x_1 and x_2 of X so that

$$\mathrm{d}_{\mathbb{H}}(z, X) = \mathrm{d}_{\mathbb{H}}(z, x_1) = \mathrm{d}_{\mathbb{H}}(z, x_2).$$

Let ℓ_{12} be the hyperbolic line segment joining x_1 to x_2, and let ℓ be the hyperbolic line containing ℓ_{12}.

By Exercise 3.18, there exists a unique point x_0 of ℓ so that $\mathrm{d}_{\mathbb{H}}(z, \ell) = \mathrm{d}_{\mathbb{H}}(z, x_0)$. Moreover, looking at the solution to Exercise 3.18, the hyperbolic distance from z to a point y of ℓ increases monotonically as a function of the hyperbolic distance $\mathrm{d}_{\mathbb{H}}(x_0, y)$ between x_0 and y.

In particular, since $\mathrm{d}_{\mathbb{H}}(z, x_1) = \mathrm{d}_{\mathbb{H}}(z, x_2)$, the point x_0 of ℓ realizing the hyperbolic distance $\mathrm{d}_{\mathbb{H}}(z, \ell)$ must lie between x_1 and x_2. That is, x_0 is contained in ℓ_{12}.

Since x_1 and x_2 are both points of X, the convexity of X gives that ℓ_{12} is contained in X, and hence that x_0 is a point of X. However, if $x_1 \neq x_2$, then

$$\mathrm{d}_{\mathbb{H}}(z, x_0) < \mathrm{d}_{\mathbb{H}}(z, x_1) = \mathrm{d}_{\mathbb{H}}(z, X),$$

which is a contradiction. This completes the proof of Proposition 5.4.

Note that open convex sets do not have the property shown to hold for closed convex sets in Proposition 5.4. For example, let U be an open half-plane determined by a hyperbolic line ℓ. Then, for each point $z \in \ell$, we have that $\mathrm{d}_{\mathbb{H}}(z, U) = 0$ but there does not exist a point $x \in U$ with $\mathrm{d}_{\mathbb{H}}(z, x) = 0$.

All of the examples of convex sets given to this point have been either open or closed subsets of the hyperbolic plane. There are also convex sets in the hyperbolic plane that are neither open nor closed.

To take a specific example, let U be an open half-plane determined by a hyperbolic line ℓ, let x and y be two points on ℓ, and let ℓ_{xy} be the closed hyperbolic line segment joining x and y. Then, the union $U \cup \ell_{xy}$ is convex, but is neither open nor closed.

There is a common way of generating convex sets in the hyperbolic plane, namely by taking convex hulls. Given a subset Y of the hyperbolic plane, the *convex hull* $\text{conv}(Y)$ of Y is the intersection of all the convex sets in the hyperbolic plane containing Y.

For example, for the set $Y = \{x, y\}$ containing two distinct points, the convex hull $\text{conv}(Y)$ of Y is equal to the closed hyperbolic line segment ℓ_{xy} joining x and y. To see this, we first recall that by Proposition 5.2, we know that ℓ_{xy} is convex.

Hence, it remains only to show that there does not exist a convex set containing x and y that is properly contained in ℓ_{xy}. But from the definition of convexity, we see that any convex set containing x and y must contain the closed hyperbolic line segment ℓ_{xy} joining them, and so $\text{conv}(Y) = \ell_{xy}$.

Naively, we should expect the convex hull of a convex set to be the convex set back again, and this is indeed the case.

Exercise 5.3

Let X be a convex set in the hyperbolic plane. Prove that $\text{conv}(X) = X$.

We saw above that the convex hull of a pair of points in the hyperbolic plane is the closed hyperbolic line segment joining the two points, and so the convex hull of a non-convex set can be considerably larger than the set.

Exercise 5.4

Let ℓ_1 and ℓ_2 be two hyperbolic lines. Determine the convex hull $\text{conv}(\ell_1 \cup \ell_2)$ of their union.

One advantage to our approach to defining convex sets, namely by defining a convex set as the intersection of a collection of half-planes, is that we can generalize the scope of our definition. Namely, we can define what it means for a subset of the union of the hyperbolic plane and its circle at infinity to be convex, without actually altering the definition in any essential way. To make this explicit, we work in the upper half-plane model $\mathbb{H}$, whose circle at infinity is $\overline{\mathbb{R}}$.

Each hyperbolic line ℓ in $\mathbb{H}$ determines a pair of points in $\overline{\mathbb{R}}$, namely its endpoints at infinity, and each half-plane determined by ℓ is naturally associated to one of the two arcs in $\overline{\mathbb{R}}$ determined by this pair of points. So, with only a slight abuse of language, we can speak of a half-plane in $\mathbb{H}$ containing a point in $\overline{\mathbb{R}}$.

In particular, for a subset X of $\mathbb{H} \cup \overline{\mathbb{R}}$, define the *convex hull* $\text{conv}(X)$ of X in $\mathbb{H}$ to be the intersection of all the half-planes in $\mathbb{H}$ containing X.

For example, if x and y are two distinct points in $\overline{\mathbb{R}}$, the convex hull $\text{conv}(Y)$ of the set $Y = \{x, y\}$ is the hyperbolic line determined by x and y. Similarly, if z is a point of $\mathbb{H}$ and if x is a point of $\overline{\mathbb{R}}$, the convex hull $\text{conv}(Z)$ of the set $Z = \{z, x\}$ is the hyperbolic ray determined by z and x.

We close this section by describing a notion that is related to convexity but that is weaker than convexity. Note that, in order to determine whether a subset Y of the hyperbolic plane is or is not convex, it is necessary to consider all the closed hyperbolic line segments joining pairs of points of Y, and to check whether each is itself contained in Y. In some sense, then, convexity is a notion without a preferred base-point.

Definition 5.5

A set Y in the hyperbolic plane is *star-like with respect to a point* $x \in Y$ if, for each $y \in Y$, the closed hyperbolic line segment ℓ_{xy} joining x to y lies in Y. A set is *star-like* if it is star-like with respect to some point.

One way to think of star-like is as convexity with a choice of base-point.

It is reasonable to view a set being star-like as a weaker notion than being convex, since all convex sets are star-like, but not all star-like sets are convex.

To see this, suppose that X is a convex set in the hyperbolic plane, and let x be any point of X. By the definition of convexity, for every point $y \in X$, the closed hyperbolic line segment ℓ_{xy} joining x to y is contained in X. Hence, we have that X is star-like with respect to x.

In fact, the choice of x is arbitrary, and so a convex set is star-like with respect to every point.

On the other hand, it is not difficult to construct a set that is star-like with respect to some point but that is not convex. In fact, we have already seen an example of such a set in this section.

Exercise 5.5

Construct a set Y in the hyperbolic plane that is star-like with respect to some point $x \in Y$ but that is not convex.

5.2 A Characterization of Convex Sets

The reason we said in Section 5.1 that half-planes are the most basic convex sets in the hyperbolic plane is that the convex sets in the hyperbolic plane are exactly the sets that can be expressed as the intersection of a collection of half-planes. The purpose of this section is to give a proof of this characterization.

Theorem 5.6

A subset X of the hyperbolic plane is convex if and only if X can be expressed as the intersection of a collection of half-planes.

We have already proven one direction of Theorem 5.6. We know from Section 5.1 that half-planes are convex, and we know from Exercise 5.1 that the intersection of a collection of convex sets is convex. Hence, the intersection of a collection of half-planes is convex.

Suppose now that X is a convex set in the hyperbolic plane. It remains only to show that X can be expressed as the intersection of a collection of half-planes. For the remainder of this section, we work in the upper half-plane model $\mathbb{H}$ of the hyperbolic plane.

In order to get a feel for the question being considered, we start with a particular example, namely the positive imaginary axis I in $\mathbb{H}$. Since I is a hyperbolic line in $\mathbb{H}$, we know from Proposition 5.2 that I is convex.

To express I as the intersection of a collection of closed half-planes, consider the two closed half-planes A and B determined by I, namely

$$A = \{z \in \mathbb{H} \mid \mathrm{Re}(z) \geq 0\} \text{ and } B = \{z \in \mathbb{H} \mid \mathrm{Re}(z) \leq 0\}$$

Then, $A \cap B = I = \{z \in \mathbb{H} \mid \mathrm{Re}(z) = 0\}$.

We may also express I as the intersection of a collection of open half-planes, by expressing each of the closed half-planes above as the intersection of a collection of open half-planes. Specifically, for each $\varepsilon > 0$, let

$$A_\varepsilon = \{z \in \mathbb{H} \mid \mathrm{Re}(z) > -\varepsilon\} \text{ and } B_\varepsilon = \{z \in \mathbb{H} \mid \mathrm{Re}(z) < \varepsilon\}.$$

Then, we can express A as the intersection $A = \cap_{\varepsilon>0} A_\varepsilon$ and B as the intersection $B = \cap_{\varepsilon>0} B_\varepsilon$. Hence, we can express I as

$$I = \cap_{\varepsilon>0}(A_\varepsilon \cap B_\varepsilon).$$

Since Möb($\mathbb{H}$) acts transitively on the set $\mathcal{L}$ of hyperbolic lines in $\mathbb{H}$, this argument shows that every hyperbolic line can be expressed both as the intersection of a collection of closed half-planes and as the intersection of a collection of open half-planes.

Exercise 5.6

Express a closed hyperbolic ray and a closed hyperbolic line segment both as the intersection of a collection of open half-planes, and as the intersection of a collection of closed half-planes.

In general, the proof that a convex set can be expressed as the intersection of a collection of half-planes follows the same general outline as the argument just given, showing that a hyperbolic line can be expressed as the intersection of a collection of half-planes.

However, the construction of this collection of half-planes for a general convex set involves a number of technical complications, some of which come from the fact that convex sets need not be open or closed. Consequently, we complete the proof only for the special case of a closed convex set X. The interested reader is encouraged to ponder the general case.

Let X be a closed convex subset of $\mathbb{H}$, and let z be a point of $\mathbb{H}$ that is not contained in X. By Proposition 5.4, there exists a unique point $x_z \in X$ with $\mathrm{d}_{\mathbb{H}}(z, x_z) = \mathrm{d}_{\mathbb{H}}(z, X)$.

Let M_z be the hyperbolic line segment joining x_z and z, and let L_z be the hyperbolic line perpendicular to M_z and passing through x_z. The hyperbolic line L_z is the bounding line for two half-planes, the open half-plane A_z containing z and the closed half-plane B_z not containing z.

We now show that X is contained in B_z. Suppose not. Since A_z and B_z are disjoint half-planes whose union is $\mathbb{H}$, there must then exist a point p_z of $X \cap A_z$. Let ℓ_z be the hyperbolic line segment joining x_z to p_z, and let ℓ be the hyperbolic line containing ℓ_z.

Since M_z is perpendicular to L_z, and since M_z and L_z intersect at x_z, we have by Proposition 5.4 that

$$\mathrm{d}_{\mathbb{H}}(z, y) \geq \mathrm{d}_{\mathbb{H}}(z, x_z)$$

for every point y in L_z, with equality if and only if $y = x_z$.

Also, for any point y of B_z that is not contained in L_z, the hyperbolic line segment joining y to z intersects L_z, and so we have that

$$\mathrm{d}_{\mathbb{H}}(z, y) > \mathrm{d}_{\mathbb{H}}(L_z, z) = \mathrm{d}_{\mathbb{H}}(x_z, z)$$

as well. That is, we have that

$$\mathrm{d}_{\mathbb{H}}(y, z) \geq \mathrm{d}_{\mathbb{H}}(x_z, z)$$

for every point y of the closed half-plane B_z, with equality if and only if $y = x_z$.

Now apply Proposition 5.4 to the point z and the hyperbolic line ℓ. The only hyperbolic line through x_z that intersects M_z perpendicularly is L_z. Since p_z is contained in A_z and since A_z and L_z are disjoint, we have that $\ell \neq L_z$, and so ℓ and M_z cannot intersect perpendicularly.

Since ℓ and M_z do not intersect perpendicularly, the solution to Exercise 3.18 implies that there exists a point a of ℓ so that

$$d_{\mathbb{H}}(a, z) < d_{\mathbb{H}}(x_z, z).$$

By the argument just given, this point a cannot lie in B_z, and so there exists a point a of $\ell \cap A_z$ so that

$$d_{\mathbb{H}}(a, z) < d_{\mathbb{H}}(x_z, z).$$

Let a_ℓ be the point of ℓ given by the solution to Exercise 3.18 that satisfies

$$d_{\mathbb{H}}(z, a_\ell) \leq d_{\mathbb{H}}(z, a)$$

for every point a of ℓ. Since $d_{\mathbb{H}}(a, z)$ is monotone increasing as a function of $d_{\mathbb{H}}(a, a_\ell)$, we see that a_ℓ is contained in A_z as well.

In particular, regardless of whether a_ℓ is or is not contained in ℓ_z, there exists a point b_z of ℓ_z that satisfies

$$d_{\mathbb{H}}(z, b_z) < d_{\mathbb{H}}(z, x_z).$$

However, since X is convex and since both of the endpoints p_z and x_z of ℓ_z are points of X, we have that ℓ_z is contained in X. This is where we make use of the convexity of X.

So, we have constructed a point b_z of X for which

$$d_{\mathbb{H}}(z, b_z) < d_{\mathbb{H}}(z, x_z).$$

This contradicts the choice of x_z. This contradiction completes the proof of the claim that X is contained in the closed half-plane B_z.

To complete the proof of Theorem 5.6, note that we can express X as the intersection

$$X = \cap\{B_z \mid z \in \mathbb{H} \text{ and } z \notin X\}.$$

5.3 Hyperbolic Polygons

As in Eulidean geometry, the *polygon* is one of the basic objects in hyperbolic geometry. In the Euclidean plane, a polygon is a closed convex set that is bounded by Euclidean line segments. We would like to mimic this definition as much as possible in the hyperbolic plane.

Starting from the definition of convexity described in Section 5.1 and characterized in Section 5.2, namely that a convex set is the intersection of a collection of half-planes, we need to impose a condition on this collection. The condition we impose is *local finiteness.*

Definition 5.7

Let $\mathcal{H} = \{H_\alpha\}_{\alpha \in A}$ be a collection of half-planes in the hyperbolic plane, and for each $\alpha \in A$, let ℓ_α be the bounding line for H_α. The collection $\mathcal{H}$ is *locally finite* if for each point z in the hyperbolic plane, there exists some $\varepsilon > 0$ so that only finitely many of the bounding lines ℓ_α intersect the open hyperbolic disc $U_\varepsilon(z)$ of hyperbolic radius ε and hyperbolic centre z.

In words, even though the collection $\{H_\alpha\}$ may be infinite, near each point it looks as though it is a finite collection when viewed in the hyperbolic disc of radius ε. Note that the value of ε needed will in general be a function of the point z.

It is easy to see that every finite collection $\mathcal{H} = \{H_k\}_{1 \le k \le n}$ of half-planes is locally finite, since every open disc $U_\varepsilon(z)$ in the hyperbolic plane can intersect at most n of the bounding lines, since there are only n half-planes in the collection.

Less easy to see is that there cannot exist an uncountable collection of half-planes that is locally finite.

Exercise 5.7

Prove that an uncountable collection of distinct half-planes in the hyperbolic plane cannot be locally finite.

One example of an infinite collection of half-planes that is locally finite is the collection $\{H_n\}_{n \in \mathbb{Z}}$ in $\mathbb{H}$, where the bounding line ℓ_n of H_n lies in the Euclidean circle with Euclidean centre n and Euclidean radius 1, and where H_n is the closed half-plane determined by ℓ_n that contains the point $2i$. Part of this collection of bounding lines is shown in Fig. 5.2

To see that $\{H_n\}$ is locally finite, take some point $x \in \mathbb{H}$. For each $\varepsilon > 0$, consider the open hyperbolic disc $U_\varepsilon(x)$ with hyperbolic centre x and hyperbolic radius ε. The hyperbolic distance between x and the hyperbolic line ℓ_μ

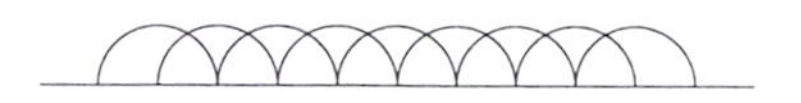

Figure 5.2: Some bounding lines

contained in the Euclidean line $\{z \in \mathbb{H} \mid \text{Re}(z) = \text{Re}(x) + \mu\}$ satisfies

$$\text{d}_{\mathbb{H}}(x, \ell_\mu) < \frac{\mu}{\text{Im}(x)},$$

since the right hand side is the hyperbolic length of the Euclidean line segment joining x to $x + \mu$.

In particular, the hyperbolic disc $U_\varepsilon(x)$ is contained in the strip

$$\{z \in \mathbb{H} \mid \text{Re}(x) - \varepsilon\, \text{Im}(x) < \text{Re}(z) < \text{Re}(x) + \varepsilon\, \text{Im}(x)\}.$$

Since for each $\varepsilon > 0$ this strip intersects only finitely many of the ℓ_n, we see that the collection $\{H_n\}$ is locally finite.

However, just because a collection of half-planes is countable does not imply that it is locally finite. For example, consider the collection $\mathcal{H} = \{H_n\}_{n \in \mathbb{N}}$ of closed half-planes in $\mathbb{H}$, where the bounding line ℓ_n of H_n is the hyperbolic line in $\mathbb{H}$ contained in the Euclidean circle of Euclidean radius 1 and Euclidean centre $\frac{1}{n}$, and where H_n is the closed half-plane determined by ℓ_n that contains $2i$.

To see that $\mathcal{H}$ is not locally finite, we observe that for each $\varepsilon > 0$ the open hyperbolic disc $U_\varepsilon(i)$ intersects infinitely many of the ℓ_n, including all of those for which the hyperbolic distance $\text{d}_{\mathbb{H}}(i, \frac{1}{n} + i)$ satisfies $\text{d}_{\mathbb{H}}(i, \frac{1}{n} + i) < \varepsilon$.

Definition 5.8

A *hyperbolic polygon* is a closed convex set in the hyperbolic plane that can be expressed as the intersection of a locally finite collection of closed half-planes.

One thing to note about this definition is that for a given hyperbolic polygon P, there will always be many different locally finite collections of closed half-planes whose intersection is P. Also, we use closed half-planes in the definition, since a closed subset of $\mathbb{H}$ cannot be expressed as the intersection of a locally finite collection of open half-planes.

We have already seen one example of a hyperbolic polygon in $\mathbb{H}$, namely $\cap_{n \in \mathbb{Z}} H_n$, where H_n is the closed half-plane determined by the hyperbolic line ℓ_n contained in the Euclidean circle with Euclidean centre $n \in \mathbb{Z}$ and Euclidean radius 1.

Another example of a hyperbolic polygon in $\mathbb{H}$ is shown in Fig. 5.3. It is the intersection of the five closed half-planes $H_1 = \{z \in \mathbb{H} \mid \text{Re}(z) \leq 1\}$, $H_2 = \{z \in \mathbb{H} \mid \text{Re}(z) \geq -1\}$, $H_3 = \{z \in \mathbb{H} \mid |z| \geq 1\}$, $H_4 = \{z \in \mathbb{H} \mid |z-1| \geq 1\}$, and $H_5 = \{z \in \mathbb{H} \mid |z+1| \geq 1\}$.

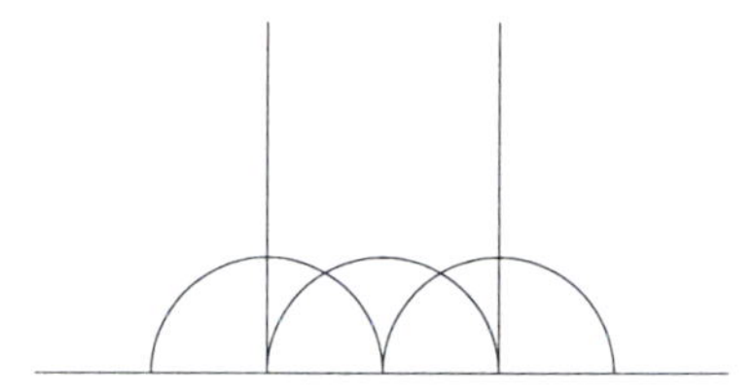

Figure 5.3: A hyperbolic polygon

In addition to individual hyperbolic polygons, we can also consider families of hyperbolic polygons. For this example we work in the Poincaré disc $\mathbb{D}$. For $r > 1$, consider the hyperbolic polygon P_r that is the intersection of the four closed half-planes

$$H_k = \{z \in \mathbb{D} \mid |z - ri^k| \geq \sqrt{r^2 - 1}\}$$

for $k = 0, 1, 2, 3$. For an illustration of such a P_r with $r = 1.5$, see Fig. 5.4.

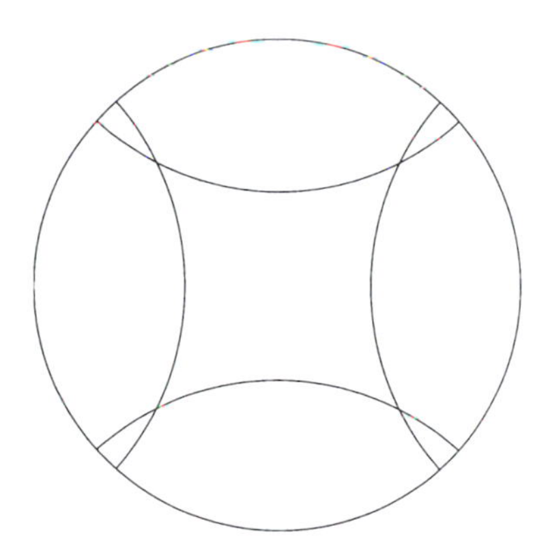

Figure 5.4: A hyperbolic polygon in $\mathbb{D}$

Up to this point, none of our definitions have made use of any intrinsic property of any specific model of the hyperbolic plane. In fact, everything we have said makes sense in every model, and so we are free to apply these definitions in whichever model is most convenient or most comfortable.

Note that by the definition of hyperbolic polygon we have chosen, there are some subsets of the hyperbolic plane that satisfy the definition of a hyperbolic polygon but that we do not want to consider as hyperbolic polygons.

For example, a hyperbolic line ℓ is a hyperbolic polygon, as it is a closed convex set in the hyperbolic plane that can be expressed as the intersection of the collection $\{A_\ell, B_\ell\}$ of closed half-planes, where A_ℓ and B_ℓ are the two closed half-planes determined by ℓ.

It is a bit uncomfortable having a hyperbolic line as a hyperbolic polygon. One way to get around this is to impose another condition. Recall that the *interior* of a set X in the hyperbolic plane is the largest open set contained in X. The interior of a hyperbolic line is empty, since a hyperbolic line does not contain an open subset of the hyperbolic plane.

Say that a hyperbolic polygon is *non-degenerate* if it has non-empty interior, and say that a hyperbolic polygon is *degenerate* if it has empty interior.

Unless explicitly stated otherwise, we assume that *all hyperbolic polygons are non-degenerate*. For instance, all the examples of hyperbolic polygons given above are non-degenerate. And as it turns out, the degenerate hyperbolic polygons are easy to understand.

Exercise 5.8

Prove that a degenerate hyperbolic polygon is either a hyperbolic line, a closed hyperbolic ray, or a closed hyperbolic line segment.

Let P be a hyperbolic polygon in the hyperbolic plane. The boundary ∂P of P has a very nice decomposition. To see this decomposition, let ℓ be a hyperbolic line that intersects P. It may be that ℓ intersects the interior of P. In this case, the intersection $P \cap \ell$ is a closed convex subset of ℓ that is not a point, and so is either a closed hyperbolic line segment in ℓ, a closed hyperbolic ray in ℓ, or all of ℓ.

On the other hand, it may be that ℓ does not pass through the interior of P. In this case, P is contained in one of the closed half-planes determined by ℓ. The proof of this is very similar to the analysis carried out in detail in Section 5.2.

The intersection $P \cap \ell$ is again a closed convex subset of ℓ, and so is either a point in ℓ, a closed hyperbolic line segment in ℓ, a closed hyperbolic ray in ℓ, or all of ℓ. All four possibilities can occur, as is shown in Fig. 5.5.

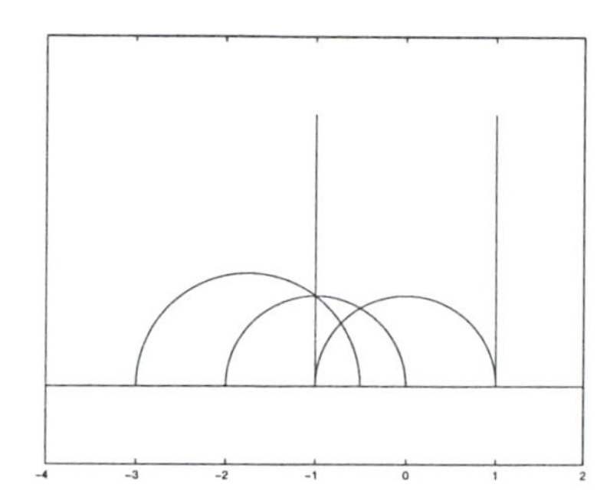

Figure 5.5: Intersections of hyperbolic lines with a hyperbolic polygon

Here, the hyperbolic polygon P is the intersection of four closed half-planes, namely $H_1 = \{z \in \mathbb{H} \mid \mathrm{Re}(z) \le 1\}$, $H_2 = \{z \in \mathbb{H} \mid \mathrm{Re}(z) \ge -1\}$, $H_3 = \{z \in \mathbb{H} \mid |z| \ge 1\}$, and $H_4 = \{z \in \mathbb{H} \mid |z+1| \ge 1\}$.

The bounding lines of P intersect P in turn in a hyperbolic line, a closed hyperbolic ray, a closed hyperbolic ray, and a closed hyperbolic line segment. The hyperbolic line ℓ whose endpoints at infinity are -3 and $-\frac{1}{2}$ intersects P in a single point.

In general, let P be a hyperbolic polygon and let ℓ be a hyperbolic line so that P intersects ℓ and so that P is contained in one of the closed half-planes determined by ℓ. If the intersection $P \cap \ell$ is a point, we say that this point is a *vertex* of P. In the other cases, namely that the intersection $P \cap \ell$ is either a closed hyperbolic line segment, a closed hyperbolic ray, or all of ℓ, we say that this intersection is a *side* of P. The sides and vertices of a hyperbolic polygon are very closely related.

Lemma 5.9

Let P be a hyperbolic polygon. The vertices of P are the endpoints of the sides of P.

Lemma 5.9 is a fairly direct consequence of our definition of a hyperbolic polygon as the intersection of a locally finite collection of closed half-planes of the hyperbolic plane.

To start the proof of Lemma 5.9, express P as the intersection of a locally finite collection $\mathcal{H}$ of distinct closed half-planes. Write $\mathcal{H} = \{H_n\}_{n \in A}$, where A is a (necessarily) countable set, and let ℓ_n be the bounding line of H_n.

Let p be a point of ∂P. The local finiteness of $\mathcal{H}$ implies that there exists some $\varepsilon_0 > 0$ so that only finitely many of the ℓ_n intersect the open hyperbolic disc $U_{\varepsilon_0}(p)$.

For $\delta < \varepsilon_0$, the number of bounding lines that intersect $U_\delta(p)$ is bounded above by the number of bounding lines that intersect $U_{\varepsilon_0}(p)$. In particular, as $\delta \to 0$, the number of bounding lines intersecting $U_\delta(p)$ either stays constant or decreases.

As there are only finitely many bounding lines that intersect $U_{\varepsilon_0}(p)$, there exists some $\varepsilon < \varepsilon_0$ so that all the bounding lines that intersect $U_\varepsilon(p)$ actually pass through p. This is the crucial point at which we make use of the local finiteness of the collection $\mathcal{H}$.

Let $H_1, \ldots, H_n$ be the closed half-planes in $\mathcal{H}$ whose bounding lines contain p, and consider their intersection. Since P is non-degenerate, P is not contained in a hyperbolic line, and so no two of these closed half-planes can have the same bounding line.

The n bounding lines break the hyperbolic disc $U_\varepsilon(p)$ into $2n$ wedge-shaped regions. The intersection $\cap_{k=1}^n H_k$ is one of these wedge-shaped regions. An illustration of this phenomenon in the Poincaré disc $\mathbb{D}$ with the vertex $p = 0$ is given in Fig. 5.6.

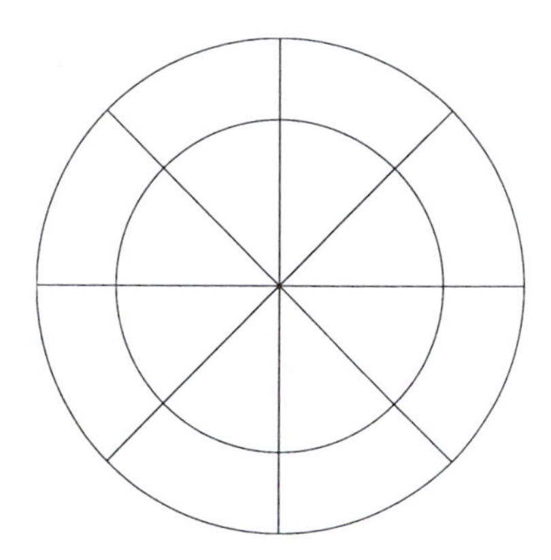

Figure 5.6: Wedges of a hyperbolic disc

Note that there are necessarily two half-planes H_j and H_m in the collection $\mathcal{H}$ so that $\cap_{k=1}^n H_k = H_j \cap H_m$.

In particular, the vertex p is the point of intersection of the two bounding lines

ℓ_j and ℓ_m, and the two sides of P that contain p are the sides of P contained in ℓ_j and ℓ_m. This completes the proof of Lemma 5.9.

The proof of Lemma 5.9 shows that there exists a very good local picture of the structure of the boundary of a hyperbolic polygon P. In fact, given a hyperbolic polygon, we can make use of this proof to construct a canonical locally finite collection of closed half-planes $\mathcal{H}$ whose intersection is P.

Namely, let P be a hyperbolic polygon in the hyperbolic plane. Construct a collection $\mathcal{H}$ of closed half-planes as follows. Enumerate the sides of P as $s_1, \ldots, s_k, \ldots$. For each s_k, let ℓ_k be the hyperbolic line that contains s_k, and let H_k be the closed half-plane determined by ℓ_k that contains P. Then, $\mathcal{H} = \{H_k\}$ is a locally finite collection of closed half-planes, and

$$P = \cap_{H \in \mathcal{H}} H.$$

One consequence of this analysis is that each vertex v of a hyperbolic polygon P is the intersection of two sides of P. In particular, we can measure the *interior angle* inside P at v.

Definition 5.10

Let P be a hyperbolic polygon, and let v be a vertex of P that is the intersection of two sides s_1 and s_2 of P. Let ℓ_k be the hyperbolic line containing s_k. The union $\ell_1 \cup \ell_2$ divides the hyperbolic plane into four components, one of which contains P. The *interior angle* of P at v is the angle between ℓ_1 and ℓ_2, measured in the component of the complement of $\ell_1 \cup \ell_2$ containing P.

We close this section by discussing some basic types of hyperbolic polygons. We first consider compact hyperbolic polygons. Since a compact hyperbolic polygon P is necessarily bounded (by the definition of compactness), and since P necessarily has only many finite sides (by the local finiteness of the collection of half-planes whose intersection is P), and since hyperbolic rays and hyperbolic lines are not bounded, all of the sides of P are closed hyperbolic line segments.

In fact, a bit more is true.

Exercise 5.9

Let P be a compact hyperbolic polygon. Prove that P is the convex hull of its vertices.

As in the Euclidean plane, there are a number of hyperbolic polygons with particular names. A *hyperbolic triangle* is a compact hyperbolic polygon with three sides. A *hyperbolic quadrilateral* is a compact hyperbolic polygon with four sides, a *hyperbolic rhombus* is a hyperbolic quadrilateral whose sides have equal length, and a *hyperbolic square* is a hyperbolic rhombus with all right angles.

More generally, a *hyperbolic n-gon* is a compact hyperbolic polygon with n sides. A hyperbolic n-gon is *regular* if all its sides have equal length and if all its interior angles are equal.

A *hyperbolic parallelogram* is a hyperbolic quadrilateral whose opposite sides are contained in parallel or ultraparallel hyperbolic lines. Note that, since parallelism is a much different condition in the hyperbolic plane than it is in the Euclidean plane, there is a much greater variety of possible hyperbolic parallelograms than there are Euclidean parallelograms. Also, as opposed to the Euclidean case, if P is a hyperbolic quadrilateral, it may be very difficult to determine whether P is a hyperbolic parallelogram or not.

For example, consider the hyperbolic quadrilateral Q in $\mathbb{H}$ with vertices $x_1 = i - 1$, $x_2 = 2i - 1$, $x_3 = i + 1$, and $x_4 = 2i + 1$. A picture of Q is given in Fig. 5.7. Let s_{jk} denote the side of Q connecting the vertices x_j and x_k, and let ℓ_{jk} be the hyperbolic line containing s_{jk}.

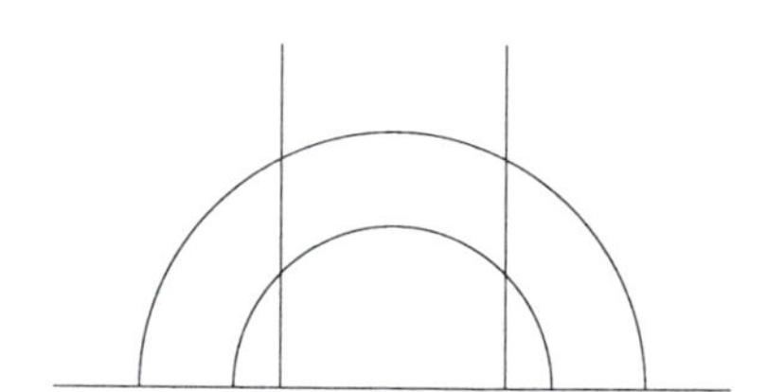

Figure 5.7: A hyperbolic quadrilateral

Since ℓ_{12} and ℓ_{34} are contained in the Euclidean lines $\{z \in \mathbb{H} \mid \text{Re}(z) = -1\}$ and $\{z \in \mathbb{H} \mid \text{Re}(z) = 1\}$, respectively, we see that s_{12} and s_{34} are contained in parallel hyperbolic lines.

Since ℓ_{13} and ℓ_{24} are contained in the Euclidean circles $\{z \in \mathbb{H} \mid |z| = \sqrt{2}\}$ and $\{z \in \mathbb{H} \mid |z| = \sqrt{5}\}$, respectively, we see that s_{13} and s_{24} are contained in parallel hyperbolic lines. Hence, Q is a hyperbolic parallelogram.

Exercise 5.10

For $s > 2$, let Q_s be the hyperbolic quadrilateral in $\mathbb{H}$ with vertices

$x_1 = i - 1$, $x_2 = 2i - 1$, $x_3 = i + 1$, and $x_4 = si + 1$. Determine the values of s for which Q_s is a hyperbolic parallelogram.

Though we have concentrated our attention up to this point on compact hyperbolic polygons, there is no requirement in the definition that hyperbolic polygons be compact. In fact, non-compact hyperbolic polygons will play an important role in the later sections.

There are several flavours of non-compactness.

Definition 5.11

Say that a hyperbolic polygon P in $\mathbb{H}$ has an *ideal vertex* at $v \in \overline{\mathbb{R}}$ if there are two sides of P that are either closed hyperbolic rays or hyperbolic lines and that share v as an endpoint at infinity.

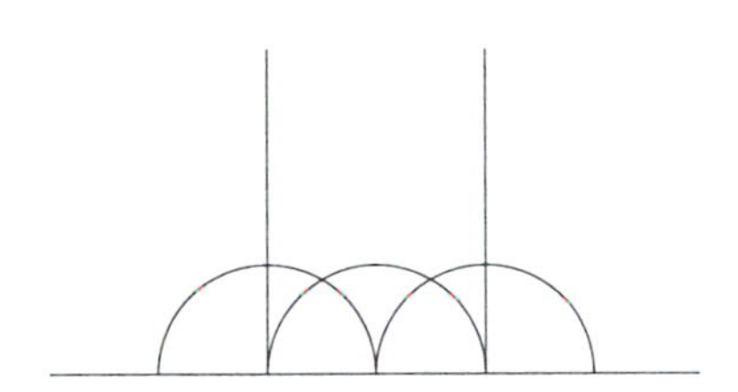

Figure 5.8: A hyperbolic polygon with an ideal vertex at ∞

For each integer $n \geq 3$, an *ideal n-gon* is a hyperbolic polygon P that has n sides and n ideal vertices. In particular, each side of an ideal polygon P is the hyperbolic line determined by a pair of ideal vertices. See Fig. 5.9 for an ideal hyperbolic triangle and a non-ideal hyperbolic triangle.

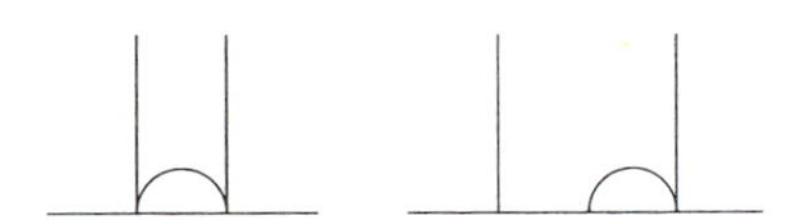

Figure 5.9: An ideal hyperbolic triangle and a non-ideal hyperbolic triangle

Exercise 5.11

Let P be an ideal polygon, and let $\{p_1, \ldots, p_k\}$ be its ideal vertices. Prove that $P = \text{conv}(\{p_1, \ldots, p_k\})$.

Exercise 5.12

Let T be a hyperbolic triangle in $\mathbb{H}$ with sides A, B, and C. For any point $x \in A$, prove that

$$\mathrm{d}_{\mathbb{H}}(x, B \cup C) \leq \ln(1 + \sqrt{2}).$$

5.4 The Definition of Hyperbolic Area

In addition to all those we have already mentioned, one of the nice properties of hyperbolic convex sets in general, and hyperbolic polygons in particular, is that it is easy to calculate their hyperbolic area. But first, we need to define hyperbolic area. For now, we work in the upper half-plane model $\mathbb{H}$.

Recall that in $\mathbb{H}$, the hyperbolic length of a piecewise differentiable path, and from this the hyperbolic distance between a pair of points, is calculated by integrating the hyperbolic element of arc-length $\frac{1}{\text{Im}(z)}|\mathrm{d}z|$ along the path. The hyperbolic area of a set X in $\mathbb{H}$ is given by integrating the square of the hyperbolic element of arc-length over the set.

Definition 5.12

The *hyperbolic area* $\text{area}_{\mathbb{H}}(X)$ of a set X in $\mathbb{H}$ is given by the integral

$$\text{area}_{\mathbb{H}}(X) = \int_X \frac{1}{\text{Im}(z)^2}\, \mathrm{d}x\, \mathrm{d}y = \int_X \frac{1}{y^2}\, \mathrm{d}x\, \mathrm{d}y,$$

where $z = x + iy$.

For example, consider the region X in $\mathbb{H}$ that is bounded by the three Euclidean lines $\{z \in \mathbb{H} \mid \text{Re}(z) = -1\}$, $\{z \in \mathbb{H} \mid \text{Re}(z) = 1\}$, and $\{z \in \mathbb{H} \mid \text{Im}(z) = 1\}$. Note that since $\{z \in \mathbb{H} \mid \text{Im}(z) = 1\}$ is not contained in a hyperbolic line, the region X is not a hyperbolic polygon.

The hyperbolic area of X is then

$$\text{area}_{\mathbb{H}}(X) = \int_X \frac{1}{y^2}\, \mathrm{d}x\, \mathrm{d}y = \int_{-1}^{1} \int_{1}^{\infty} \frac{1}{y^2}\, \mathrm{d}y\, \mathrm{d}x = \int_{-1}^{1} \mathrm{d}x = 2.$$

Exercise 5.13

For $s > 0$, let X_s be the region in $\mathbb{H}$ bounded by the three Euclidean lines $\{z \in \mathbb{H} \,|\, \mathrm{Re}(z) = -1\}$, $\{z \in \mathbb{H} \,|\, \mathrm{Re}(z) = 1\}$, and $\{z \in \mathbb{H} \,|\, \mathrm{Im}(z) = s\}$. Calculate the hyperbolic area $\mathrm{area}_{\mathbb{H}}(X)$ of X_s.

In our discussion of hyperbolic lengths of piecewise differentiable paths, we actually derived the hyperbolic element of arc-length under the assumption that it was invariant under Möb($\mathbb{H}$). It then followed immediately that hyperbolic length was naturally invariant under Möb($\mathbb{H}$).

However, as we will see in Exercise 5.15, we cannot derive the formula for hyperbolic area by assuming invariance under the action of Möb($\mathbb{H}$), as the group of transformations of $\mathbb{H}$ preserving hyperbolic area is much larger than Möb($\mathbb{H}$). So, we spend the remainder of this section giving a direct proof that hyperbolic area is invariant under the action of Möb($\mathbb{H}$) = Isom($\mathbb{H}$).

Theorem 5.13

Hyperbolic area in $\mathbb{H}$ is invariant under the action of Möb($\mathbb{H}$). That is, if X be a set in $\mathbb{H}$ whose hyperbolic area $\mathrm{area}_{\mathbb{H}}(X)$ is defined and if A is an element of Möb($\mathbb{H}$), then

$$\mathrm{area}_{\mathbb{H}}(X) = \mathrm{area}_{\mathbb{H}}(A(X)).$$

The proof of Theorem 5.13 is an application of the change of variables theorem from multivariable calculus, which we recall here. Let $F : \mathbb{R}^2 \to \mathbb{R}^2$ be a differentiable function, which we write as

$$F(x, y) = (f(x, y), g(x, y)),$$

and consider its derivative DF, written in matrix form as

$$DF(x,y) = \begin{pmatrix} \frac{\partial f}{\partial x}(x,y) & \frac{\partial g}{\partial x}(x,y) \\ \frac{\partial f}{\partial y}(x,y) & \frac{\partial g}{\partial y}(x,y) \end{pmatrix}.$$

The change of variables theorem states that, under fairly mild conditions on a set X in $\mathbb{R}^2$ and a function h on X, we have

$$\int_{F(X)} h(x,y)\, \mathrm{d}x\, \mathrm{d}y = \int_X h \circ F(x,y)\, |\det(DF)|\, \mathrm{d}x\, \mathrm{d}y.$$

We do not give the most general statement of the conditions for the change of variables theorem. For our purposes, it suffices to note that the change of variables theorem applies to convex subsets X of $\mathbb{H}$ and to continuous functions h.

We begin by applying the change of variables theorem to an element A of $\text{Möb}^+(\mathbb{H})$. We first rewrite A in terms of x and y as

$$\begin{aligned} A(z) = \frac{az+b}{cz+d} &= \frac{(az+b)(c\overline{z}+d)}{(cz+d)(c\overline{z}+d)} \\ &= \frac{acx^2+acy^2+bd+bcx+adx+iy}{(cx+d)^2+c^2y^2}, \end{aligned}$$

where a, b, c, $d \in \mathbb{R}$ and $ad - bc = 1$.

So, consider the function $A : \mathbb{H} \to \mathbb{H}$ given by

$$A(x,y) = \left(\frac{acx^2+acy^2+bd+bcx+adx}{(cx+d)^2+c^2y^2}, \frac{y}{(cx+d)^2+c^2y^2} \right).$$

Calculating, we see that

$$DA(x,y) = \begin{pmatrix} \frac{(cx+d)^2-c^2y^2}{((cx+d)^2+c^2y^2)^2} & \frac{2cy(cx+d)}{((cx+d)^2+c^2y^2)^2} \\ \frac{-2cy(cx+d)}{((cx+d)^2+c^2y^2)^2} & \frac{(cx+d)^2-c^2y^2}{((cx+d)^2+c^2y^2)^2} \end{pmatrix}.$$

In particular, we have that

$$\det(DA(x,y)) = \frac{1}{((cx+d)^2+c^2y^2)^2}.$$

For the calculation of hyperbolic area in $\mathbb{H}$, we are integrating the function $h(x,y) = \frac{1}{y^2}$, and so we also need to calculate the composition

$$h \circ A(x,y) = \frac{((cx+d)^2+c^2y^2)^2}{y^2}.$$

Hence, the change of variables theorem yields that

$$\begin{aligned} \text{area}_{\mathbb{H}}(A(X)) &= \int_{A(X)} \frac{1}{y^2}\, dx\, dy \\ &= \int_X h \circ A(x,y)\, |\det(DA)|\, dx\, dy \\ &= \int_X \frac{((cx+d)^2+c^2y^2)^2}{y^2} \frac{1}{((cx+d)^2+c^2y^2)^2}\, dx\, dy \\ &= \int_X \frac{1}{y^2}\, dx\, dy = \text{area}_{\mathbb{H}}(X), \end{aligned}$$

as desired.

In order to complete the proof of Theorem 5.13, we need only show that hyperbolic area is invariant under $B(z) = -\overline{z}$, which is the content of the following exercise.

Exercise 5.14

Use the change of variables theorem to prove that hyperbolic area in $\mathbb{H}$ is invariant under $B(z) = -\overline{z}$.

As mentioned earlier in this section, unlike in the case of hyperbolic lengths, in which Möb($\mathbb{H}$) is exactly the group of transformations of $\mathbb{H}$ preserving hyperbolic length, there are transformations of $\mathbb{H}$ that preserve hyperbolic area but that do not lie in Möb($\mathbb{H}$).

Exercise 5.15

Consider the homeomorphism f of $\mathbb{H}$ given by $f(z) = z + \text{Im}(z)$. Use the change of variables theorem to prove that f preserves hyperbolic area. Show further that f is not an element of Möb($\mathbb{H}$).

Though we do not prove it, we note that this definition of hyperbolic area makes sense for every convex set in $\mathbb{H}$, and for many non-convex sets as well. We do not address the general question of determining the sets in $\mathbb{H}$ for which this definition of hyperbolic area makes sense.

5.5 Area and the Gauss-Bonnet Formula

Now that we have shown that hyperbolic area in $\mathbb{H}$ is invariant under the action of Möb($\mathbb{H}$), we are more easily able to calculate the hyperbolic area of relatively simple sets in the hyperbolic plane, such as hyperbolic polygons. We begin by considering hyperbolic triangles.

One approach to proceed by direct calculation. That is, for a hyperbolic triangle P, we write down explicit expressions for the Euclidean lines and Euclidean circles containing the sides of P, and use these as the limits of integration to calculate the hyperbolic area of P.

Even for a specific hyperbolic triangle, this approach is not very effective, as we see in Exercise 5.16. And to derive the formula for a general hyperbolic triangle, this approach is far too unwieldly.

Exercise 5.16

Consider the hyperbolic triangle P in $\mathbb{H}$ with vertices $v_1 = i$, $v_2 = 2+2i$, and $v_3 = 4 + i$. Write down the integral giving the hyperbolic area of P.

Another approach to try is to express our given hyperbolic triangle somehow in terms of hyperbolic triangles whose hyperbolic areas are significantly easier to calculate. This is the approach we take.

We begin with a simple example. Consider a hyperbolic triangle P with one ideal vertex v_1, and with two other vertices v_2 and v_3, which might or might not be ideal vertices. Let ℓ_{jk} be the hyperbolic line determined by v_j and v_k.

We now make use of the transitivity properties of Möb($\mathbb{H}$) as described in Section 2.9. Namely, let γ be an element of Möb($\mathbb{H}$) that takes v_1 to ∞ and that takes ℓ_{23} to the hyperbolic line contained in the unit circle, so that $v_2 = e^{i\varphi}$ and $v_3 = e^{i\theta}$, where $0 \leq \theta < \varphi \leq \pi$. (We allow $\theta = 0$ and $\varphi = \pi$ to allow for the possibility that one or both of v_2 and v_3 is an ideal vertex.) See Fig. 5.10.

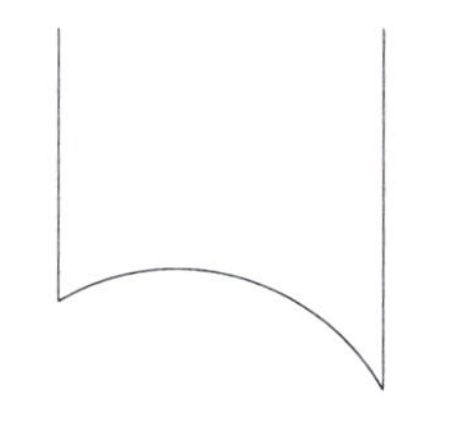

Figure 5.10: The case of one ideal vertex

Since hyperbolic area is invariant under the action of Möb($\mathbb{H}$), we may thus assume P to be the hyperbolic triangle with an ideal vertex at ∞, and with two other vertices at $e^{i\varphi}$ and $e^{i\theta}$, where $0 \leq \theta < \varphi \leq \pi$. Since P has at least one ideal vertex, it is not compact, but we can still easily calculate its hyperbolic area.

Calculating, we see that

$$\text{area}_{\mathbb{H}}(P) = \int_P \frac{1}{y^2}\,\mathrm{d}x\,\mathrm{d}y = \int_{\cos(\varphi)}^{\cos(\theta)} \int_{\sqrt{1-x^2}}^{\infty} \frac{1}{y^2}\mathrm{d}y\,\mathrm{d}x = \int_{\cos(\varphi)}^{\cos(\theta)} \frac{1}{\sqrt{1-x^2}}\mathrm{d}x.$$

Making the substitution $x = \cos(w)$, so that $\mathrm{d}x = -\sin(w)\,\mathrm{d}w$, this becomes

$$\int_{\cos(\varphi)}^{\cos(\theta)} \frac{1}{\sqrt{1-x^2}}\mathrm{d}x = \int_{\varphi}^{\theta} -\mathrm{d}w = \varphi - \theta.$$

At this point, with a hint of foreshadowing, we observe that the interior angle of P at the ideal vertex $v_1 = \infty$ is $\alpha_1 = 0$, the interior angle at the vertex $v_2 = e^{i\theta}$ is $\alpha_2 = \theta$, and the interior angle at the vertex $v_3 = e^{i\varphi}$ is $\alpha_3 = \pi - \varphi$. Hence, we have proven the following proposition.

Proposition 5.14

Let P be a hyperbolic triangle with one ideal vertex, and let α_2 and α_3 be the interior angles at the other two vertices, which might or might not be ideal vertices. Then,

$$\text{area}_{\mathbb{H}}(P) = \pi - (\alpha_2 + \alpha_3).$$

One consequence of Proposition 5.14 is that the hyperbolic area of an ideal triangle in $\mathbb{H}$ is π. This follows from the observation that the interior angle at each ideal vertex of an ideal triangle is 0.

Suppose now that P is a compact hyperbolic triangle with vertices v_1, v_2, and v_3. Let α_k be the interior angle of P at v_k. Let ℓ be the hyperbolic ray from v_1 passing through v_2 and let x be the endpoint at infinity of ℓ. See Fig. 5.11.

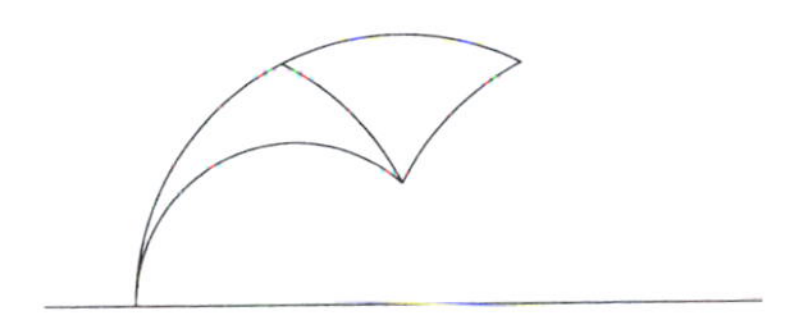

Figure 5.11: The case of no ideal vertices

The hyperbolic triangle T with vertices v_1, v_3, and x has one ideal vertex at x and two non-ideal vertices at v_1 and v_3. The interior angle of T at v_1 is α_1 and the interior angle of T at v_3 is $\delta > \alpha_3$. So, by Proposition 5.14 the hyperbolic area of T is

$$\text{area}_{\mathbb{H}}(T) = \pi - (\alpha_1 + \delta).$$

The hyperbolic triangle T' with vertices v_2, v_3, and x has one ideal vertex at x and two non-ideal vertices at v_2 and v_3. The interior angle of T' at v_2 is $\pi - \alpha_2$, and the interior angle of T' at v_3 is $\delta - \alpha_3$. So, the hyperbolic area of T' is

$$\text{area}_{\mathbb{H}}(T') = \pi - (\pi - \alpha_2 + \delta - \alpha_3).$$

Since T is the union of T' and P, and since T' and P overlap only along a side, we see that

$$\text{area}_{\mathbb{H}}(T) = \text{area}_{\mathbb{H}}(T') + \text{area}_{\mathbb{H}}(P).$$

Substituting in the calculations of the previous two paragraphs, we see that

$$\begin{aligned}
\text{area}_{\mathbb{H}}(P) &= \text{area}_{\mathbb{H}}(T) - \text{area}_{\mathbb{H}}(T') \\
&= \pi - (\alpha_1 + \delta) - (\pi - (\pi - \alpha_2 + \delta - \alpha_3)) \\
&= \pi - (\alpha_1 + \alpha_2 + \alpha_3).
\end{aligned}$$

This completes the proof of the following theorem.

Theorem 5.15

Let P be a hyperbolic triangle with interior angles α, β, and γ. Then,

$$\text{area}_{\mathbb{H}}(P) = \pi - (\alpha + \beta + \gamma).$$

Theorem 5.15 is known as the *Gauss-Bonnet formula.* Looking back at our calculations for hyperbolic triangles with ideal vertices, since the interior angle at an ideal vertex is 0, we see that this formula holds as well for hyperbolic triangles with ideal vertices and not only for compact hyperbolic triangles.

Exercise 5.17

Consider the hyperbolic triangle P in $\mathbb{H}$ given in Exercise 5.16, with vertices i, $4+i$, and $2+2i$. Calculate the hyperbolic area of P by determining the three interior angles of P.

We can generalize Theorem 5.15 to general hyperbolic polygons. In order to do so, we need to specify the class of hyperbolic polygons we wish to consider.

For any finite sided hyperbolic polygon P, the sum of the numbers of vertices and ideal vertices of P is at most the number of sides of P, since each side of P contains at most two vertices or ideal vertices of P. We work only with finite sided hyperbolic polygons for which the sum of the numbers of vertices and ideal vertices is equal to the number of sides.

Note that there are hyperbolic polygons for which the sum of the numbers of vertices and ideal vertices is less than the number of sides. For example, a closed half-plane has one side but no vertices and no ideal vertices.

Theorem 5.16

Let P be a finite sided hyperbolic polygon with vertices and ideal vertices $v_1, \ldots, v_n$, and assume that the sum of the numbers of vertices and ideal vertices is equal to the number of sides. Let α_k be the interior angle at v_k. Then,

$$\text{area}_{\mathbb{H}}(P) = (n-2)\pi - \sum_{k=1}^{n} \alpha_k.$$

We prove Theorem 5.16 by decomposing P into a number of hyperbolic triangles, using Theorem 5.15 to calculate the hyperbolic area of each hyperbolic triangle in this decomposition, and then summing to get the hyperbolic area of P.

Choose a point x in the interior of P. Since P is convex, the hyperbolic line segment (or hyperbolic ray, in the case that v_k is an ideal vertex) ℓ_k joining x to v_k is contained in P. The hyperbolic line segments $\ell_1, \ldots, \ell_n$ break P up into n triangles $T_1, \ldots, T_n$. See Fig. 5.12.

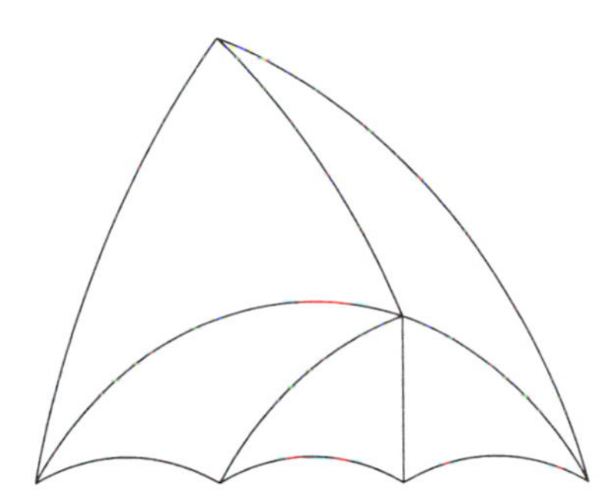

Figure 5.12: Decomposing a hyperbolic pentagon into hyperbolic triangles

Label these hyperbolic triangles so that T_k has vertices x, v_k, and v_{k+1} for $1 \leq k \leq n$, where in a slight abuse of notation we set $v_{n+1} = v_1$ and $T_{n+1} = T_1$.

Let μ_k be the interior angle of T_k at x, and note that

$$\sum_{k=1}^{n} \mu_k = 2\pi.$$

Let β_k be the interior angle of T_k at v_k and let δ_k be the interior angle of T_k at v_{k+1}. Since both T_k and T_{k+1} have a vertex at v_{k+1}, we see that

$$\alpha_{k+1} = \delta_k + \beta_{k+1}.$$

Applying Theorem 5.15 to T_k yields that

$$\text{area}_{\mathbb{H}}(T_k) = \pi - (\mu_k + \beta_{k+1} + \delta_k).$$

Since the union $T_1 \cup \cdots \cup T_n$ is equal to P and since the hyperbolic triangles $T_1, \ldots, T_n$ overlap only along on their sides, we have that

$$\begin{aligned} \text{area}_{\mathbb{H}}(P) &= \sum_{k=1}^{n} \text{area}_{\mathbb{H}}(T_k) = \sum_{k=1}^{n} [\pi - (\mu_k + \beta_{k+1} + \delta_k)] \\ &= n\pi - \left[\sum_{k=1}^{n} \mu_k + \sum_{k=1}^{n} \beta_{k+1} + \sum_{k=1}^{n} \delta_k \right]. \end{aligned}$$

Since $\alpha_{k+1} = \delta_k + \beta_{k+1}$ for each k, we have that

$$\sum_{k=1}^{n} \beta_{k+1} + \sum_{k=1}^{n} \delta_k = \sum_{k=1}^{n} \alpha_k.$$

Hence,

$$\text{area}_{\mathbb{H}}(P) = \sum_{k=1}^{n} \text{area}_{\mathbb{H}}(T_k) = (n-2)\pi - \sum_{k=1}^{n} \alpha_k.$$

This completes the proof of Theorem 5.16.

Note that we could have taken the point x around which we decomposed P to be a point on a side of P, or a vertex of P. In either case, the particulars of the calculation would be slightly different, but we would still obtain Theorem 5.16 in the end.

In the former case, we would decompose P into $n-1$ hyperbolic triangles, and the sum of the interior angles of the hyperbolic triangles at x would equal π.

In the latter case, we would decompose P into $n-2$ hyperbolic triangles, and the sum of the interior angles of the hyperbolic triangles at x would equal the interior angle of P at x.

In the same way that we defined the hyperbolic element of area in $\mathbb{H}$, this entire discussion can be carried out in any of the other models of the hyperbolic plane, such as the Poincaré disc model $\mathbb{D}$. In particular, Theorem 5.16 holds in any model of the hyperbolic plane.

In the Poincaré disc $\mathbb{D}$, there are two natural coordinate systems that come from the fact that $\mathbb{D}$ is a subset of $\mathbb{C}$, namely the standard cartesian coordinates and polar coordinates. In the cartesian coordinates x and y, the hyperbolic area of a set X in $\mathbb{D}$ is written

$$\text{area}_{\mathbb{D}}(X) = \int_X \frac{4}{(1-|z|^2)^2}\, dx\, dy = \int_X \frac{4}{(1-x^2-y^2)^2}\, dx\, dy.$$

In polar coordinates, using the standard conversion from cartesian to polar coordinates $x = r\cos(\theta)$ and $y = r\sin(\theta)$, this integral becomes

$$\text{area}_{\mathbb{D}}(X) = \int_X \frac{4r}{(1-r^2)^2}\,\mathrm{d}r\,\mathrm{d}\theta.$$

Exercise 5.18

Given $s > 0$, let D_s be the open hyperbolic disc in $\mathbb{D}$ with hyperbolic centre 0 and hyperbolic radius s. Show that the hyperbolic area $\text{area}_{\mathbb{D}}(D_s)$ of D_s is

$$\text{area}_{\mathbb{D}}(D_s) = 4\pi \sinh^2\left(\frac{1}{2}s\right).$$

Exercise 5.19

Let D_s be as defined in Exercise 5.18. Describe the behaviour of the quantity

$$q_{\mathbb{D}}(s) = \frac{\text{length}_{\mathbb{D}}(S_s)}{\text{area}_{\mathbb{D}}(D_s)}.$$

Compare the behaviour of $q_{\mathbb{D}}(s)$ with the corresponding quantity $q_{\mathbb{C}}$ calculated using a Euclidean circle and a Euclidean disc in $\mathbb{C}$.

We close this section with the following observation. Though we do not give a proof of it, there is a general formula relating the hyperbolic length $\text{length}_{\mathbb{D}}(C)$ of a Jordan curve C in $\mathbb{D}$ and the hyperbolic area $\text{area}_{\mathbb{D}}(D)$ of the region D in $\mathbb{D}$ bounded by C. Specifically,

$$[\text{length}_{\mathbb{D}}(C)]^2 - 4\pi\,\text{area}_{\mathbb{D}}(D) - [\text{area}_{\mathbb{D}}(D)]^2 \geq 0.$$

Such an inequality, which is not specific to the hyperbolic plane, is called an *isoperimetric inequality*, as it can be viewed as an equation describing the maximum area of all regions bounded by Jordan curves of a fixed length.

Note that in $\mathbb{D}$, it follows from Exercise 4.4 and Exercise 5.18 that the minimum of 0 is achieved when C is a hyperbolic circle and D is the hyperbolic disc bounded by C. However, we are not able to conclude from this that any region in $\mathbb{D}$ that achieves the minimum of 0 in this isoperimetric inequality is in fact a hyperbolic disc.

For more information about isoperimetric inequalities in general, the interested reader is referred to the encyclopedic work of Burago and Zalgaller [5] and the references contained therein.

5.6 Applications of the Gauss-Bonnet Formula

In this section, we describe two applications of Theorem 5.16 in the hyperbolic plane. One is positive, in that it asserts the existence of a large number and variety of different regular compact hyperbolic polygons. The other is negative, in that it asserts the non-existence of a certain type of transformation of the hyperbolic plane.

We begin with a fact about the Euclidean plane $\mathbb{C}$, namely that for each integer $n \geq 3$, there exists only one regular Euclidean n-gon, up to scaling, rotation, and translation.

Here is one construction of a regular Euclidean n-gon P_n in $\mathbb{C}$. Start by choosing a basepoint x in $\mathbb{C}$, and let $\ell_1, \ldots, \ell_n$ be n Euclidean rays from x, where the angle between consecutive rays is $\frac{2\pi}{n}$. Choose some $r > 0$, and for each k consider the point y_k on ℓ_k that is Euclidean distance r from x. These points $y_1, \ldots, y_n$ are the vertices of a regular Eulidean n-gon P_n.

To see that P_n is unique up to scaling, rotation, and translation, we repeat the construction. That is, choose a different basepoint x' in $\mathbb{C}$. Let $\ell'_1, \ldots, \ell'_n$ be n Euclidean rays from x', where the angle between consecutive rays is $\frac{2\pi}{n}$. Choose some $r' > 0$, and let y'_k be the point on ℓ'_k that is Euclidean distance r' from x'. Then, the points $y'_1, \ldots, y'_n$ are the vertices of a regular Euclidean n-gon P'_n.

We now construct a transformation of $\mathbb{C}$ that takes P_n to P'_n. Let θ be the angle between ℓ_1 and the positive real axis, and let θ' be the angle between ℓ'_1 and the positive real axis. Then, the homeomorphism B of $\mathbb{C}$ given by

$$B(z) = e^{i(\theta'-\theta)} \, \frac{r'}{r} \, (z - x + x')$$

is the composition of a rotation, a dilation, and a translation of $\mathbb{C}$ that satisfies $B(P_n) = P'_n$.

In particular, the interior angles of P_n at its vertices depend only on the number of sides n, and not on the choice of the basepoint x or the Euclidean rays ℓ_k or the Euclidean distance r of the vertices of P_n from x. In fact, the interior angle at a vertex of P_n is $\frac{n-2}{n}\pi$.

In the hyperbolic plane, the situation is considerably different.

Proposition 5.17

For each $n \geq 3$ and for each α in the interval $(0, \frac{n-2}{n}\pi)$, there is a compact regular hyperbolic n-gon whose interior angle is α.

We work in the Poincaré disc $\mathbb{D}$, and start with the same construction just given for regular Euclidean n-gons in $\mathbb{C}$. Given $n \geq 3$, consider the n hyperbolic rays $\ell_0, \ldots, \ell_{n-1}$ from 0, where ℓ_k is the hyperbolic ray determined by 0 and $p_k = \exp\left(\frac{2\pi i}{n}k\right)$.

For each $0 < r < 1$, the n points $rp_0 = r, \ldots, rp_{n-1} = r\exp\left(\frac{2\pi i}{n}(n-1)\right)$ in $\mathbb{D}$ are the vertices of a regular hyperbolic n-gon $P_n(r)$.

We first show that $P_n(r)$ is a hyperbolic polygon. We do this by expressing $P_n(r)$ as the intersection of a locally finite collection of closed half-planes. For $0 \leq k \leq n-1$, let ℓ_k be the hyperbolic line passing through p_k and p_{k+1}, where we again engage in a slight abuse of notation and set $p_n = p_0$.

Let H_k be the closed half-plane determined by ℓ_k that contains 0. The fact that $P_n(r)$ is a hyperbolic polygon follows immediately from the observation that

$$P_n(r) = \cap_{k=0}^{n-1} H_k.$$

To see that $P_n(r)$ is regular, we may use the elliptic Möbius transformation

$$m(z) = \exp\left(\frac{2\pi i}{n}\right) z,$$

which is contained in $\text{Möb}(\mathbb{D})$. For each $1 \leq k \leq n-1$, we have that $m^k(rp_0) = rp_k$ and that $m^k(\ell_0) = \ell_k$.

The sides of $P_n(r)$ that intersect at rp_0 lie in the two hyperbolic lines ℓ_{n-1} and ℓ_0. Since $m^k(\ell_0) = \ell_k$ and $m^k(\ell_{n-1}) = \ell_{k-1}$, we also have that m^k takes the two sides of $P_n(r)$ that intersect at rp_0 to the two sides of $P_n(r)$ that intersect at rp_k.

In particular, the interior angles of $P_n(r)$ at any two vertices are equal, and the hyperbolic lengths of any two sides of $P_n(r)$ are equal.

For $0 < r < 1$, let $\alpha(r)$ denote the interior angle of $P_n(r)$ at $r = rp_0$. We now analyze the behaviour of $\alpha(r)$ as r varies. We note that $\alpha(r)$ is a continuous function of r, by the calculation in Exercise 5.20.

Exercise 5.20

Express the interior angle of $P_n(r)$ at $r = rp_0$ in terms of n and r. Conclude that $\alpha(r)$ is a continuous function of r.

Theorem 5.16 yields that the hyperbolic area of $P_n(r)$ is

$$\text{area}_{\mathbb{D}}(P_n(r)) = (n-2)\pi - \sum_{k=0}^{n-1} \alpha(r) = (n-2)\pi - n\alpha(r).$$

For each value of $0 < r < 1$, the hyperbolic polygon $P_n(r)$ is contained in the hyperbolic disc D_r in $\mathbb{D}$ with hyperbolic centre 0 and Euclidean radius r. (Note that this implies the compactness of $P_n(r)$.) The hyperbolic area of D_r is

$$\text{area}_{\mathbb{D}}(D_r) = \frac{2\pi r}{1 - r^2}.$$

Since $P_n(r)$ is contained in D_r, we have that $\text{area}_{\mathbb{D}}(P_n(r)) \leq \text{area}_{\mathbb{D}}(D_r)$, and so

$$\lim_{r \to 0^+} \text{area}_{\mathbb{D}}(P_n(r)) \leq \lim_{r \to 0^+} \frac{2\pi r}{1 - r^2} = 0.$$

Substituting in the expression for $\text{area}_{\mathbb{D}}(P_n(r))$, we see that

$$\lim_{r \to 0^+} [(n-2)\pi - n\alpha(r)] = 0,$$

and so

$$\lim_{r \to 0^+} \alpha(r) = \frac{n-2}{n}\pi.$$

As r increases, there are two observations we can make, both of which we can get either from Exercise 5.20 or from direct observation.

First, for $0 < s < r < 1$, the vertices of $P_n(s)$ lie in the interior of $P_n(r)$. The convexity of $P_n(r)$ then forces $P_n(s)$ to be contained in $P_n(r)$, and so

$$\text{area}_{\mathbb{D}}(P_n(s)) < \text{area}_{\mathbb{D}}(P_n(r))$$

for $0 < s < r < 1$. In words, the hyperbolic area of $P_n(r)$ is monotonically increasing in r. Since

$$\text{area}_{\mathbb{D}}(P_n(r)) = (n-2)\pi - n\alpha(r),$$

we have that the interior angle $\alpha(r)$ is monotonically decreasing in r.

Second, as $r \to 1^-$, the compact hyperbolic polygon $P_n(r)$ is becoming more and more like the ideal hyperbolic n-gon P_n^∞ with ideal vertices at $p_0 = 1$, $p_1 = \exp(\frac{2\pi i}{n}), \ldots, \ldots, p_{n-1} = \exp(\frac{2\pi i}{n}(n-1))$. In particular, we have that

$$\lim_{r \to 1^-} \text{area}_{\mathbb{D}}(P_n(r)) = \text{area}_{\mathbb{D}}(P_n^\infty).$$

Expressing $\text{area}_{\mathbb{D}}(P_n(r))$ and $\text{area}_{\mathbb{D}}(P_n^\infty)$ in terms of the interior angles of $P_n(r)$ and P_n^∞, respectively, we have that

$$\lim_{r \to 1^-} [(n-2)\pi - n\alpha(r)] = (n-2)\pi,$$

and so

$$\lim_{r \to 1^-} \alpha(r) = 0.$$

Combining these observations, we see that for $n \geq 3$, the interior angle $\alpha(r)$ of the compact regular hyperbolic n-gon $P_n(r)$ lies in the interval $(0, \frac{n-2}{n}\pi)$. Moreover, the monotonicity and continuity of α imply that every number in this interval is the interior angle of one and only one of the hyperbolic polygons $P_n(r)$. This completes the proof of Proposition 5.17.

One specific way in which the behaviour of hyperbolic polygons is much different from the behaviour of Euclidean polygons is that, in the Euclidean plane, there is one and only one regular n-gon with all right angles, namely the square.

However, in the hyperbolic plane, not only do hyperbolic squares not exist, but for each $n \geq 5$ there exists a compact regular hyperbolic n-gon with all right angles.

To see that hyperbolic squares do not exist, we use Proposition 5.17 in the case $n = 4$. The interval of possible interior angles of a compact regular hyperbolic 4-gon is $(0, \frac{1}{2}\pi)$. In particular, there is no hyperbolic 4-gon with all right angles.

The proof that there exist compact regular hyperbolic n-gons with all right angles for $n \geq 5$ is left as an exercise.

Exercise 5.21

Prove that for $n \geq 5$, there exists a compact regular hyperbolic n-gon all of whose interior angles are right angles.

In addition to the fact that the interior angle $\alpha(r)$ of the compact regular hyperbolic n-gon $P_n(r)$ is a continuous function of r, we also have that the hyperbolic length of a side of $P_n(r)$ is continuous as a function of r.

Exercise 5.22

Given $0 < r < 1$, explicitly calculate the hyperbolic length of a side of $P_n(r)$ in terms of n and r.

For each $n \geq 5$, Exercise 5.21 gives one compact hyperbolic n-gon with all right angles, namely the compact regular hyperbolic n-gon with all right angles. In fact, for each $n \geq 5$ there are many non-regular compact hyperbolic n-gons with all right angles, though we do not prove this fact here.

Also, it is possible to construct hyperbolic polygons with prescribed interior angles that are not necessarily right angles. In fact, the only restriction on the possible internal angles is that the hyperbolic area, as given by the Gauss-Bonnet formula, should be positive. Again, we do not prove this fact here. The interested reader is referred to Beardon [2] for the proof of Theorem 5.18.

Theorem 5.18

(Beardon [2]) Let $\alpha_1, \ldots, \alpha_n$ be a collection of n numbers in the open interval $(0, \pi)$. Then, there exists a hyperbolic polygon in the hyperbolic plane with interior angles $\alpha_1, \ldots, \alpha_n$ if and only if

$$\alpha_1 + \cdots + \alpha_n < (n-2)\pi.$$

There is a second application of Theorem 5.16 we consider here. Recall that in the construction of regular Euclidean n-gons given at the beginning of this section, we remarked that even though the Euclidean n-gons constructed were of different area, any two regular Euclidean n-gons are related by a homeomorphism of $\mathbb{C}$ that is the composition of an isometry of $\mathbb{C}$ and a dilation of $\mathbb{C}$.

A *dilation* of $\mathbb{C}$ is a conformal homeomorphism of $\mathbb{C}$ that takes Euclidean lines to Euclidean lines. Dilations are not isometries, as they do not preserve Euclidean length or area. In fact, every dilation of $\mathbb{C}$ is of the form $f(z) = az + b$ for some $a \in \mathbb{C} - \{0\}$ and $b \in \mathbb{C}$. The map f is a Euclidean isometry if and only if $|a| = 1$, a fact which is essentially contained in the solution to Exercise 3.13.

Definition 5.19

A *hyperbolic dilation* is a conformal homeomorphism of the hyperbolic plane that takes hyperbolic lines to hyperbolic lines.

As in the case of $\mathbb{C}$, every isometry of the hyperbolic plane is a hyperbolic dilation. However, unlike in the case of $\mathbb{C}$, there are no hyperbolic dilations other than the hyperbolic isometries.

The key fact in the proof of this fact, stated as Proposition 5.20, is that since a hyperbolic dilation is conformal and so preserves angles, Theorem 5.16 immediately implies that a hyperbolic dilation preserves the hyperbolic area of a hyperbolic polygon. In particular, if g is a dilation of the hyperbolic plane and if P is a hyperbolic polygon, then the hyperbolic areas of P and of $g(P)$ are equal.

Proposition 5.20

Let f be a hyperbolic dilation of the hyperbolic plane. Then, f is a hyperbolic isometry.

The proof of Proposition 5.20 is very similar in spirit to the proof of Theorem 3.12. The main technical tool used in the proof of Proposition 5.20 is Theorem 5.16.

We work in the Poincaré disc model $\mathbb{D}$ of the hyperbolic plane. Let f be a hyperbolic dilation of $\mathbb{D}$, so that by definition f is a homeomorphism of $\mathbb{D}$ that takes hyperbolic lines to hyperbolic lines and that preserves angles.

We begin by using the transitivity properties of Möb($\mathbb{D}$) to normalize f. First compose f with an element m of Möb($\mathbb{D}$) that takes $f(0)$ to 0. By definition, every element of Möb($\mathbb{D}$) is a hyperbolic isometry, and hence is a hyperbolic dilation. Hence, the composition $m \circ f$ is a hyperbolic dilation of $\mathbb{D}$ that fixes 0.

The hyperbolic dilation $m \circ f$ of $\mathbb{D}$ takes hyperbolic rays from 0 to hyperbolic rays from 0 and preserves angles between hyperbolic rays. So, there exists an element n of Möb($\mathbb{D}$) fixing 0 so that the composition $n \circ m \circ f$ is a hyperbolic dilation of $\mathbb{D}$ that fixes 0 and that takes every hyperbolic ray from 0 to itself. (This element n will be either an elliptic Möbius transformation fixing 0, or the composition of an elliptic Möbius transformation fixing 0 and $C(z) = \overline{z}$.)

Set $g = n \circ m \circ f$. To complete the proof that f is an element of Möb($\mathbb{D}$), we show that g is the identity.

Let z_0 be a point of $\mathbb{D} - \{0\}$. Let ℓ_0 be the hyperbolic ray from 0 passing through z_0. Let ℓ_1 be the hyperbolic ray from 0 making angle $\frac{2\pi}{3}$ with ℓ_0, and let ℓ_2 be the hyperbolic ray from 0 making angle $\frac{4\pi}{3}$ with ℓ_0.

Let T be the hyperbolic triangle with vertices $v_0 = z_0$, $v_1 = \exp\left(\frac{2\pi i}{3}\right) z_0$, and $v_2 = \exp\left(\frac{4\pi i}{3}\right) z_0$. Let s_{jk} be the side of T joining v_j to v_k. We consider the image $g(T)$ of T under g.

Set $r = |z_0|$ and $s = |g(z_0)|$, so that $g(z_0) = \frac{s}{r} z_0$. We first show that $g(v_1) = \frac{s}{r} \exp\left(\frac{2\pi i}{3}\right) z_0$ and that $g(v_2) = \frac{s}{r} \exp\left(\frac{4\pi i}{3}\right) z_0$. By our assumptions on g we have that v_k, and hence $g(v_k)$, lies on the hyperbolic ray ℓ_k, since g takes each hyperbolic ray from 0 to itself. Hence, $g(v_k)$ is a positive real multiple of v_k.

Since the angle of intersection of s_{0k} with ℓ_0 is equal to the angle of intersection of s_{0k} with ℓ_k, we have that the angle of intersection of $g(s_{0k})$ with $g(\ell_0) = \ell_0$ is equal to the angle of intersection of $g(s_{0k})$ with $g(\ell_k) = \ell_k$.

In particular, the point of intersection of $g(s_{0k})$ with ℓ_0 and the point of intersection of $g(s_{0k})$ with ℓ_k are the same Euclidean distance from the origin. Since the point of intersection of $g(s_{0k})$ with ℓ_0 is $g(v_0) = \frac{s}{r} z_0$, we have that $g(v_k) = \frac{s}{r} \exp\left(\frac{2k\pi i}{3}\right) z_0$ for $k = 1$ and 2, as desired.

So, the image $g(T)$ of T under g is the hyperbolic triangle with vertices $g(v_0) = \frac{s}{r} z_0$, $g(v_1) = \frac{s}{r} \exp\left(\frac{2\pi i}{3}\right) z_0$, and $g(v_2) = \frac{s}{r} \exp\left(\frac{4\pi i}{3}\right) z_0$. Since g is a hyperbolic dilation, angles between hyperbolic lines are preserved by g, and so the interior angles of T and of $g(T)$ are equal. By Theorem 5.16, we then have that

$$\text{area}_{\mathbb{D}}(T) = \text{area}_{\mathbb{D}}(g(T)).$$

However, if $s = |g(z_0)| > r = |z_0|$, then T is properly contained in $g(T)$ and so $\text{area}_{\mathbb{D}}(T) < \text{area}_{\mathbb{D}}(g(T))$, a contradiction. If $s = |g(z_0)| < r < |z_0|$, then $g(T)$ is properly contained in T, and so $\text{area}_{\mathbb{D}}(T) > \text{area}_{\mathbb{D}}(g(T))$, which is again a contradiction.

Hence, we have that $g(z) = z$ for every point z of $\mathbb{D}$, and so g is the identity. This completes the proof of Proposition 5.20.

5.7 Trigonometry in the Hyperbolic Plane

Let T be a compact hyperbolic triangle in the hyperbolic plane. As in the case for a Euclidean triangle, there are trigonometric laws in the hyperbolic plane relating the interior angles of T and the hyperbolic lengths of the sides of T.

The way we derive the trigonometric laws in the hyperbolic plane is to link the Euclidean and hyperbolic distances between a pair of points. Since the hyperbolic and Euclidean measurement of the angles of T are the same, we may then make use of the Euclidean trigonometric laws.

As we saw in Exercise 4.2, the relationship between Euclidean and hyperbolic length involves the use of the hyperbolic trigonometric functions. Before going any further, we state some identities involving the hyperbolic trigonometric functions that arise over the course of the section, leaving their verification as an exercise.

Exercise 5.23

Verify each of the following identities.

(a) $\cosh^2(x) - \sinh^2(x) = 1$;

(b) $2\cosh(x)\sinh(x) = \sinh(2x)$;

(c) $\sinh^2(x) = \frac{1}{2}\cosh(2x) - \frac{1}{2}$;

(d) $\cosh^2(x) = \frac{1}{2}\cosh(2x) + \frac{1}{2}$;

(e) $\sinh^2(x)\cosh^2(y) + \cosh^2(x)\sinh^2(y) = \frac{1}{2}(\cosh(2x)\cosh(2y) - 1)$.

We work in the Poincaré disc model $\mathbb{D}$. Let T be a compact hyperbolic triangle in $\mathbb{D}$ with vertices v_1, v_2, and v_3. Let a, b, and c be the hyperbolic lengths of its sides, and let α, β, and γ be its interior angles, where α is the interior angle at the vertex v_1 opposite the side of hyperbolic length a, β is the interior angle at the vertex v_2 opposite the side of hyperbolic length b, and γ is the interior angle at the vertex v_3 opposite the side of hyperbolic length c.

Since the interior angles at the vertices of T and the hyperbolic lengths of the sides of T are invariant under the action of Möb($\mathbb{D}$), we may use the transitivity properties of Möb($\mathbb{D}$) to assume that $v_1 = 0$, that $v_2 = r > 0$ lies on the positive real axis, and that $v_3 = se^{i\alpha}$, where $0 < \alpha < \pi$. By Exercise 4.2, we have that

$$r = \tanh\left(\frac{1}{2}c\right) \text{ and } s = \tanh\left(\frac{1}{2}b\right).$$

On the one hand, we may apply the Euclidean law of cosines to the Euclidean triangle with vertices v_1, v_2, and v_3 to see that

$$\begin{aligned}
&|v_2 - v_3|^2 \\
&\quad = r^2 + s^2 - 2rs\cos(\alpha) \\
&\quad = \tanh^2\left(\frac{1}{2}c\right) + \tanh^2\left(\frac{1}{2}b\right) - 2\tanh\left(\frac{1}{2}c\right)\tanh\left(\frac{1}{2}b\right)\cos(\alpha).
\end{aligned}$$

On the other hand, by Proposition 4.3, we have that

$$\begin{aligned}
\frac{|v_2 - v_3|^2}{(1 - |v_2|^2)(1 - |v_3|^2)} &= \frac{|v_2 - v_3|^2}{(1 - r^2)(1 - s^2)} \\
&= \sinh^2\left(\frac{1}{2}\mathrm{d}_{\mathbb{D}}(v_2, v_3)\right) = \sinh^2\left(\frac{1}{2}a\right),
\end{aligned}$$

and so

$$\begin{aligned}
|v_2 - v_3|^2 &= (1 - r^2)(1 - s^2)\sinh^2\left(\frac{1}{2}a\right) \\
&= \operatorname{sech}^2\left(\frac{1}{2}c\right)\operatorname{sech}^2\left(\frac{1}{2}c\right)\sinh^2\left(\frac{1}{2}a\right).
\end{aligned}$$

Equating the two expressions for $|v_2 - v_3|^2$, we obtain

$$\begin{aligned}
&\operatorname{sech}^2\left(\frac{1}{2}c\right)\operatorname{sech}^2\left(\frac{1}{2}b\right)\sinh^2\left(\frac{1}{2}a\right) = \\
&\qquad \tanh^2\left(\frac{1}{2}c\right) + \tanh^2\left(\frac{1}{2}b\right) - 2\tanh\left(\frac{1}{2}c\right)\tanh\left(\frac{1}{2}b\right)\cos(\alpha).
\end{aligned}$$

And now we simplify. Multiplying through by $\cosh^2\left(\frac{1}{2}c\right)\cosh^2\left(\frac{1}{2}b\right)$, we obtain

$$\sinh^2\left(\frac{1}{2}a\right) =$$

$$\sinh^2\left(\frac{1}{2}c\right)\cosh^2\left(\frac{1}{2}b\right)+\sinh^2\left(\frac{1}{2}b\right)\cosh^2\left(\frac{1}{2}c\right)$$
$$-2\sinh\left(\frac{1}{2}c\right)\sinh\left(\frac{1}{2}b\right)\cosh\left(\frac{1}{2}c\right)\cosh\left(\frac{1}{2}b\right)\cos(\alpha).$$

Using the identities given in Exercise 5.23, this becomes

$$\frac{1}{2}\cosh(a)-\frac{1}{2}=\frac{1}{2}\cosh(b)\cosh(c)-\frac{1}{2}-\frac{1}{2}\sinh(c)\sinh(b)\cos(\alpha),$$

and so we obtain the hyperbolic **law of cosines I:**

$$\cosh(a)=\cosh(b)\cosh(c)-\sinh(c)\sinh(b)\cos(\alpha).$$

Unlike in the Euclidean plane, there are three basic trigonometric laws in the hyperbolic plane. One is the law of cosines I, which we have just derived. The other two, the hyperbolic law of sines and the hyperbolic law of cosines II, are stated below.

law of sines:

$$\frac{\sinh(a)}{\sin(\alpha)}=\frac{\sinh(b)}{\sin(\beta)}=\frac{\sinh(c)}{\sin(\gamma)}.$$

law of cosines II:

$$\cos(\gamma)=-\cos(\alpha)\cos(\beta)+\sin(\alpha)\sin(\beta)\cosh(c).$$

The hyperbolic law of cosines I and the hyperbolic law of sines are the direct analogues of the Euclidean law of cosines and the Euclidean law of sines. In fact, as we have just seen, the proof of the law of cosines I follows fairly quickly from the Euclidean law of cosines and some algebraic manipulation.

In much the same way that the Euclidean law of sines can be derived from the Euclidean law of cosines by algebraic manipulation, the hyperbolic law of sines and the hyperbolic law of cosines II can be derived from the hyperbolic law of cosines I.

Exercise 5.24

Derive the hyperbolic law of cosines II and the hyperbolic law of sines from the hyperbolic law of cosines I.

Exercise 5.25

State and prove the hyperbolic Pythagorean theorem, relating the hyperbolic lengths of the sides of a hyperbolic right triangle.

The most surprising of the hyperbolic trigonometric laws is the law of cosines II, which states that the hyperbolic length of a side of a hyperbolic triangle is determined by the interior angles of the triangle. In particular, this implies that there is a canonical unit of hyperbolic length, which is very much unlike length in the Euclidean plane.

For example, consider the compact hyperbolic triangle T with interior angles $\alpha = \frac{1}{2}\pi$, $\beta = \frac{1}{3}\pi$, and $\gamma = \frac{1}{7}\pi$ at its vertices. Let a be the hyperbolic length of the side of T opposite the vertex with angle α, let b be the hyperbolic length of the side of T opposite the vertex with angle β, and let c be the hyperbolic length of the side of T opposite the vertex with angle γ.

By the law of cosines II, the hyperbolic lengths of the three sides of T satisfy

$$\cosh(a) = \frac{\cos(\alpha) + \cos(\beta)\cos(\gamma)}{\sin(\beta)\sin(\gamma)} = \cot\left(\frac{\pi}{3}\right)\cot\left(\frac{\pi}{7}\right) \sim 1.1989;$$

$$\cosh(b) = \frac{\cos(\beta) + \cos(\alpha)\cos(\gamma)}{\sin(\alpha)\sin(\gamma)} = \cos\left(\frac{\pi}{3}\right)\csc\left(\frac{\pi}{7}\right) \sim 1.1524;$$

$$\cosh(c) = \frac{\cos(\gamma) + \cos(\alpha)\cos(\beta)}{\sin(\alpha)\sin(\beta)} = \cos\left(\frac{\pi}{7}\right)\csc\left(\frac{\pi}{3}\right) \sim 1.0404.$$

We pause to insert a note about actually solving for the hyperbolic lengths. To solve

$$\cosh(a) = \frac{1}{2}\left(e^a + e^{-a}\right) = x,$$

we see by the quadratic formula that e^a satisfies

$$e^a = x \pm \sqrt{x^2 - 1},$$

and so either

$$a = \log(x + \sqrt{x^2 - 1}) \text{ or } a = \log(x - \sqrt{x^2 - 1}).$$

However, since

$$(x + \sqrt{x^2 - 1})(x - \sqrt{x^2 - 1}) = 1,$$

we have that

$$\log(x - \sqrt{x^2 - 1}) = -\log(x + \sqrt{x^2 - 1}).$$

Since hyperbolic length is positive, we have that

$$a = \log(x + \sqrt{x^2 - 1}).$$

The hyperbolic law of cosines II has no Euclidean analogue, and in fact is false in Euclidean geometry. Indeed, one reason that the interior angles of a Euclidean triangle cannot determine the side lengths is that Euclidean geometry admits dilations. Since hyperbolic geometry does not admit dilations, as we have seen in Section 5.6, it is not unreasonable to have expected a result like the law of cosines II to hold in the hyperbolic plane.

Though we will not take this approach, we mention here that there is a unified proof of the three hyperbolic trigonometric laws, as might be suggested by the similarity of the forms of the hyperbolic laws of cosines I and II. We refer the interested reader to Thurston [27] for this approach, and for much more.

As we have seen on several occasions, including in the derivation of the hyperbolic law of cosines I, the calculation of the hyperbolic distance between points in $\mathbb{D}$ is fairly easy. However, as we have also seen on several occasions, such as in Exercise 5.20, calculations of angles in $\mathbb{D}$ can in general be very tedious.

One application of the hyperbolic trigonometric laws is to make these calculations of angle much more tractible. For instance, we may rework Exercises 5.22 and 5.20 using the two hyperbolic laws of cosines.

For $n \geq 3$ and $r > 0$, we consider the compact regular hyperbolic n-gon $P_n(r)$ in the Poincaré disc $\mathbb{D}$ with vertices at $p_k = r\exp\left(\frac{2\pi i}{n}k\right)$ for $0 \leq k \leq n-1$, as constructed in Section 5.6.

Let T be the hyperbolic triangle with vertices at 0, $p_0 = r$, and $p_1 = r\exp\left(\frac{2\pi i}{n}\right)$. The interior angle of T at 0 is $\frac{2\pi}{n}$. Also, the hyperbolic lengths of the two sides of T adjacent to 0 are equal to the hyperbolic distance from 0 to $p_0 = r$, which is

$$b = \mathrm{d}_{\mathbb{D}}(0, p_0) = \ln\left[\frac{1+r}{1-r}\right].$$

In particular,

$$\cosh(b) = \frac{1+r^2}{1-r^2} \text{ and } \sinh(b) = \frac{2r}{1-r^2}.$$

By the hyperbolic law of cosines I, the hyperbolic length a of the side of T opposite 0 satisfies

$$\cosh(a) = \cosh^2(b) - \sinh^2(b)\cos\left(\frac{2\pi}{n}\right) = \frac{(1+r^2)^2 - 4r^2\cos\left(\frac{2\pi}{n}\right)}{(1-r^2)^2}.$$

Now that we have an explicit formula for the hyperbolic length a of the side of T opposite 0, we can use the hyperbolic law of sines to determine the interior angle β of T at p_0, namely

$$\sin(\beta) = \frac{\sinh(b)\sin\left(\frac{2\pi}{n}\right)}{\sinh(a)}.$$

The interior angle of $P_n(r)$ at p_0 is then 2β.

Exercise 5.26

Let T be a compact hyperbolic triangle, all of whose sides have hyperbolic length a. Prove that the three interior angles of T are equal. Further, if we let α be the interior angle of T at a vertex, prove that

$$2\cosh\left(\frac{1}{2}a\right)\sin\left(\frac{1}{2}\alpha\right) = 1.$$

6
Groups Acting on $\mathbb{H}$

This last chapter is a very brief introduction to the study of a particularly nice class of subgroups of $\text{Möb}(\mathbb{H})$, the *discrete subgroups*. After defining *discreteness*, we describe a means of constructing a *picture of a discrete group*, namely its *fundamental polygon*. We close by giving a special case of *Poincaré's polygon theorem*, which gives conditions for a hyperbolic polygon to be the fundamental polygon of a discrete group.

6.1 The Geometry of the Action of $\text{Möb}(\mathbb{H})$

We have spent most of our time to this point studying the geometry of the hyperbolic plane, and we have made much use of the whole of its group of isometries. We now specialize to considering how individual elements act, with an eye towards considering the action of particularly nice subgroups of the group of isometries. This section is perhaps best viewed as a catalogue of possibilities.

We work for the time being in the upper half-plane model $\mathbb{H}$ of the hyperbolic plane, whose group of isometries is $\text{Isom}(\mathbb{H}) = \text{Möb}(\mathbb{H})$. As before, all the results derived in this section also hold in the other models of the hyperbolic plane, such as the Poincaré disc $\mathbb{D}$.

In Section 2.8, we saw that every non-trivial element of $\text{Möb}(\mathbb{H})$ can be written either as

$$m(z) = \frac{az+b}{cz+d}, \text{ where } a, b, c, d \text{ are real with } ad - bc = 1,$$

or as

$$n(z) = \frac{\alpha\overline{z}+\beta}{\gamma\overline{z}+\delta}, \text{ where } \alpha, \beta, \gamma, \delta \text{ are purely imaginary with } \alpha\delta - \beta\gamma = 1.$$

Using these explicit formulae, we can determine the sets of fixed points.

We first consider the case that that $m(z) = \frac{az+b}{cz+d}$ where a, b, c, and d are real with $ad - bc = 1$. What follows is very similar in spirit to the discussion in Section 2.4. In Section 2.1, we saw that the fixed points of m are the solutions to $m(z) = \frac{az+b}{cz+d} = z$, which are the roots in $\overline{\mathbb{C}}$ of the polynomial $p(z) = cz^2 + (d-a)z - b = 0$.

In the case that $c = 0$, there is one fixed point at ∞. There is a second fixed point, namely $\frac{b}{d-a}$, if and only if $a \neq d$, and such a fixed point is necessarily a real number. So, if $c = 0$, either there is a single fixed point at ∞ or there are two fixed points, one at ∞ and the other in $\mathbb{R}$.

In the case that $c \neq 0$, there are two roots of $p(z)$ in $\mathbb{C}$, namely $\frac{1}{2}[a - d \pm \sqrt{(d-a)^2 - 4bc}]$. Since the coefficients of $p(z)$ are real, the roots of $p(z)$ are invariant under complex conjugation, and so either both roots are real, or one lies in $\mathbb{H}$ and the other in the lower half-plane.

Note that $p(z)$ has exactly one root, which is then necessarily real, if and only if $(a-d)^2 - 4bc = (a+d)^2 - 4 = 0$; has two real roots if and only if $(a-d)^2 - 4bc = (a+d)^2 - 4 > 0$; and has two complex roots, symmetric under complex conjugation, if and only if $(a-d)^2 - 4bc = (a+d)^2 - 4 < 0$.

Combining this analysis with the classification of elements of Möb^+ as described in Section 2.4, we see that m has one fixed point inside $\mathbb{H}$ if and only if m is elliptic; that m has one fixed point on $\overline{\mathbb{R}}$ if and only if m is parabolic; that m has two fixed points on $\overline{\mathbb{R}}$ if and only if m is loxodromic; and that these are the only possibilities.

In the case that m is elliptic and so has one fixed point inside $\mathbb{H}$, the action of m on $\mathbb{H}$ is rotation about the fixed point. In fact, if we take the fixed point of m in $\mathbb{H}$ to be i, so that the other fixed point of m is at $-i$, we may use Exercise 2.27 to see that m has the form

$$m(z) = \frac{\cos(\theta)z + \sin(\theta)}{-\sin(\theta)z + \cos(\theta)}$$

for some real number θ. Since $\text{Möb}(\mathbb{H})$ acts transitively on $\mathbb{H}$, every elliptic element is conjugate to a Möbius transformation of this form.

Note however that m is not the standard Euclidean rotation about i. For instance, take $\theta = \frac{1}{2}\pi$ and note that $m(1+i) = -\frac{1}{2} + \frac{1}{2}i$. Indeed, the hyperbolic line passing through i and $1+i$ is not the horizontal Euclidean line $L = \{z \in \mathbb{H} \mid \text{Im}(z) = 1\}$ through i, which is not a hyperbolic line at all, but instead is the hyperbolic line contained in the Euclidean circle with Euclidean centre $\frac{1}{2}$ and Euclidean radius $\frac{\sqrt{5}}{2}$.

In the case that m is parabolic and so has one fixed point x on $\overline{\mathbb{R}}$, we may use the transitivity of Möb($\mathbb{H}$) on $\overline{\mathbb{R}}$ to conjugate m by an element of Möb($\mathbb{H}$) to have the form $m(z) = z + 1$.

In particular, a parabolic transformation m in Möb($\mathbb{H}$) with fixed point x preserves every circle in $\overline{\mathbb{C}}$ that is contained in $\mathbb{H} \cup \overline{\mathbb{R}}$ and that is tangent to $\overline{\mathbb{R}}$ at x. This is most easily seen in the case for the fixed point $x = \infty$, in which case these circles in $\overline{\mathbb{C}}$ are precisely the circles in $\overline{\mathbb{C}}$ that are the union of a horizontal Euclidean line in $\mathbb{H}$ with $\{\infty\}$. These circles are the *horocircles* invariant under m. The components of the complement of a horocircle in $\mathbb{H}$ are the two *horodiscs* in $\mathbb{H}$ determined by the horocircle.

In the case that m is loxodromic and so has two fixed points x and y in $\overline{\mathbb{R}}$, we may use the transitivity of Möb($\mathbb{H}$) on pairs of distinct points of $\overline{\mathbb{R}}$ to conjugate m to have the form $m(z) = \lambda z$ for some positive real number λ. In this case, the positive imaginary axis is taken to itself by m, and also both of the half-planes determined by the positive imaginary axis are taken to themselves by m.

In general, we define the *axis* of a loxodromic m, denoted axis(m), to be the hyperbolic line in $\mathbb{H}$ determined by the fixed points of m. Exactly as in the previous paragraph, we have that m takes its axis to itself, and also takes each of the half-planes determined by axis(m) to itself.

We summarize this analysis in the following theorem.

Theorem 6.1

Let $m(z) = \frac{az+b}{cz+d}$ be an element of Möb$^+$($\mathbb{H}$), so that a, b, c, $d \in \mathbb{R}$ and $ad - bc = 1$. Then, exactly one of the following holds:

(a) m is the identity;

(b) m has exactly two fixed points in $\overline{\mathbb{R}}$, in which case m is loxodromic and is conjugate in Möb($\mathbb{H}$) to $q(z) = \lambda z$ for some positive real number λ;

(c) m has one fixed point in $\overline{\mathbb{R}}$, in which case m is parabolic and is conjugate in Möb($\mathbb{H}$) to $q(z) = z + 1$; or

(d) m has one fixed point in $\mathbb{H}$, in which case m is elliptic and is conjugate in Möb($\mathbb{H}$) to $q(z) = \frac{\cos(\theta)z+\sin(\theta)}{-\sin(\theta)z+\cos(\theta)}$ for some real number θ.

Let m be a loxodromic transformation in $\text{Möb}^+(\mathbb{H})$, let x and y be the fixed points of m in $\overline{\mathbb{R}}$, and let A be any circle in $\overline{\mathbb{C}}$ that passes through x and y, not necessarily perpendicular to $\overline{\mathbb{R}}$. Since m preserves angles, we see that m takes $A \cap \mathbb{H}$ to itself. Further, we can see that m acts on $A \cap \mathbb{H}$ by translation.

Definition 6.2

The *translation distance* of m along $A \cap \mathbb{H}$ is $d_{\mathbb{H}}(a, m(a))$, where a is a point of $A \cap \mathbb{H}$.

In the case that $A \cap \mathbb{H}$ is equal to the axis of m, which occurs in the case that A is perpendicular to $\overline{\mathbb{R}}$, we have already calculated the translation distance of m along $A \cap \mathbb{H}$ to be

$$d_{\mathbb{H}}(\mu i, m(\mu i)) = d_{\mathbb{H}}(\mu i, \lambda \mu i) = \ln\left[\frac{\lambda\mu}{\mu}\right] = \ln(\lambda),$$

where $m(z)$ is conjugate to $q(z) = \lambda z$.

Exercise 6.1

For $\lambda > 1$, consider the loxodromic transformation $m(z) = \lambda z$. Let A be the Eulidean ray in $\mathbb{H}$ from 0 making angle θ with the positive real axis. Calculate the translation distance of m along A as a function of λ and θ.

This completes our brief tour of the action of the elements of $\text{Möb}^+(\mathbb{H})$ on $\mathbb{H}$. There are also the elements of $\text{Möb}(\mathbb{H}) - \text{Möb}^+(\mathbb{H})$ to consider, where

$$\text{Möb}(\mathbb{H}) - \text{Möb}^+(\mathbb{H}) = \{m \in \text{Möb}(\mathbb{H}) \mid m \notin \text{Möb}^+(\mathbb{H})\}.$$

As shown in Section 2.8, every element n of $\text{Möb}(\mathbb{H}) - \text{Möb}^+(\mathbb{H})$ has the form

$$n(z) = \frac{\alpha\overline{z} + \beta}{\gamma\overline{z} + \delta},$$

where α, β, γ, and δ are purely imaginary with $\alpha\delta - \beta\gamma = 1$.

As above, we begin our description of the action of n on $\mathbb{H}$ by determining the fixed points of n, which are the points z of $\mathbb{H}$ satisfying

$$\frac{\alpha\overline{z} + \beta}{\gamma\overline{z} + \delta} = z.$$

We begin our analysis by considering a particular example, namely the transformation

$$q(z) = \frac{i\overline{z} + 2i}{i\overline{z} + i}.$$

The fixed points in $\overline{\mathbb{C}}$ of q are the solutions in $\overline{\mathbb{C}}$ of $q(z) = z$, which are those points z in $\overline{\mathbb{C}}$ satisfying

$$i\overline{z} + 2i = z(i\overline{z} + i),$$

which we may rewrite as

$$-2\,\mathrm{Im}(z) + i[|z|^2 - 2] = 0.$$

Taking real and imaginary parts, we see that $\mathrm{Im}(z) = 0$ for every fixed point z of q, and so there are no fixed points of q in $\mathbb{H}$. Since $|z|^2 = 2$ as well, there are two fixed points of q in $\overline{\mathbb{R}}$, namely at $\pm\sqrt{2}$.

In this case, we see that q takes the hyperbolic line ℓ determined by $\pm\sqrt{2}$ to itself, but does not fix any point on ℓ. Instead, q acts as reflection in ℓ followed by translation along ℓ. In particular, the action of q interchanges the two half-planes in $\mathbb{H}$ determined by ℓ. We refer to q as a *glide reflection* along ℓ.

Exercise 6.2

Express q as the composition of the reflection in ℓ and a loxodromic with axis ℓ.

To attack the general case, write $\alpha = ai$, $\beta = bi$, $\gamma = ci$, and $\delta = di$, where a, b, c, and d are real with $ad - bc = -1$. Also write $x = \mathrm{Re}(z)$ and $y = \mathrm{Im}(z)$. The equation for the fixed points of n then becomes

$$c|z|^2 + dz - a\overline{z} - b = cx^2 + cy^2 + (d-a)x - b + i(d+a)y = 0.$$

Assume that n has a fixed point $z = x + iy$ in $\mathbb{H}$. As the imaginary part of the fixed point must be non-zero, we see that $a + d = 0$, and so $d = -a$. In particular, we see that $ad - bc = -d^2 - bc = -1$. The fixed points of n are then given by the equation

$$cx^2 + cy^2 + 2dx - b = 0.$$

In the case that $c = 0$, we have no restriction on the imaginary part of the fixed point z. Also, we have that $d \neq 0$ since $ad - bc = -1$, and so the fixed points of n are exactly the points in $\mathbb{H}$ that lie on the Euclidean line $\{z \in \mathbb{H} | \mathrm{Re}(z) = \frac{b}{2d}\}$, which is the hyperbolic line determined by ∞ and $\frac{b}{2d}$.

That is, every point on the the hyperbolic line ℓ determined by ∞ and $\frac{b}{2d}$ is fixed by n. It follows from the discussion in Section 3.6 that an isometry of $\mathbb{H}$ that fixes every point on a hyperbolic line is either the identity or is reflection in that hyperbolic line. In particular, this yields that n is reflection in ℓ.

Exercise 6.3

Determine the fixed points of $q(z) = -\overline{z} + 1$.

In the case that $c \neq 0$, divide through by c and complete the square to see that the fixed points of n in $\mathbb{H}$ are given by the equation

$$x^2 + y^2 + \frac{2d}{c}x - \frac{b}{c} = \left(x + \frac{d}{c}\right)^2 + y^2 - \frac{d^2 + bc}{c^2} = \left(x + \frac{d}{c}\right)^2 + y^2 - \frac{1}{c^2} = 0,$$

which is the Euclidean circle A in $\mathbb{C}$ with Euclidean centre $-\frac{d}{c}$ and Euclidean radius $\frac{1}{|c|}$.

In particular, this gives that the fixed points of n are exactly the points on the hyperbolic line $A \cap \mathbb{H}$. As in the case that $c = 0$, in this case n is equal to reflection in $A \cap \mathbb{H}$.

Exercise 6.4

Determine the fixed points of

$$q(z) = \frac{2i\overline{z} - i}{3i\overline{z} - 2i}.$$

We need to exercise a bit of caution, however, as there are elements of Möb($\mathbb{H}$), such as the transformation $q(z) = \frac{i\overline{z}+2i}{i\overline{z}+i}$ considered earlier in the section, which do not act as reflection in a hyperbolic line.

The difficulty lies in the fact that we began the analysis of the elements of Möb($\mathbb{H}$) $-$ Möb$^+$($\mathbb{H}$) by assuming that the element in question had a fixed point in $\mathbb{H}$.

So, to order to complete our analysis of the elements of Möb($\mathbb{H}$) $-$ Möb$^+$($\mathbb{H}$), we consider the case in which there are no fixed points of n in $\mathbb{H}$.

In this case, the solutions of $n(z) = z$ are the points z in $\overline{\mathbb{C}}$ that satisfy the equation

$$cx^2 + cy^2 + (d - a)x - b + i(d + a)y = 0.$$

Since we are interested in the case that there are no solutions in $\mathbb{H}$, we set $y = 0$ and consider those solutions that lie in $\overline{\mathbb{R}}$.

In the case that $c = 0$, we have two solutions, namely ∞ and $\frac{b}{2d}$. In this case, n takes the hyperbolic line ℓ determined by ∞ and $\frac{b}{2d}$ to itself and interchanges the two half-planes determined by ℓ, but no point on ℓ is fixed by n, since n

has no fixed points in $\mathbb{H}$ by assumption. That is, n acts as a glide reflection along ℓ.

In this case, we can express n as the composition of reflection in ℓ and a loxodromic with axis ℓ. The easiest way to see this is to note that the composition

$$n \circ B(z) = n(-\overline{z}) = \frac{-\alpha z + \beta}{-\gamma z + \delta} = \frac{-az + b}{-cz + d}$$

is loxodromic, where $\alpha = ai$, $\beta = bi$, $\gamma = ci$, and $\delta = di$ are purely imaginary with $\alpha\delta - \beta\gamma = 1$.

In the case that $c \neq 0$, the fixed points of n can be found by applying the quadratic equation to $cx^2 + (d-a)x - b = 0$, to get

$$x = \frac{1}{2c}\left[a - d \pm \sqrt{(d-a)^2 + 4bc}\right] = \frac{1}{2c}\left[a - d \pm \sqrt{(a+d)^2 + 4}\right],$$

using that $ad - bc = -1$.

In particular, if n has no fixed points in $\mathbb{H}$, then n necessarily has exactly two fixed points on $\overline{\mathbb{R}}$, and so n acts as a glide reflection along the hyperbolic line determined by these two points. Exactly as in the cases above, such an n is the composition of reflection in this hyperbolic line and a loxodromic with this hyperbolic line as its axis.

We summarize the analysis of elements of $\text{Möb}(\mathbb{H}) - \text{Möb}^+(\mathbb{H})$ in the following theorem.

Theorem 6.3

Let $n(z) = \frac{\alpha\overline{z}+\beta}{\gamma\overline{z}+\delta}$ be an element of $\text{Möb}(\mathbb{H}) - \text{Möb}^+(\mathbb{H})$, so that α, β, γ, and δ are purely imaginary with $\alpha\delta - \beta\gamma = 1$. Then, exactly one of the following holds:

1 n fixes a point of $\mathbb{H}$, in which case there is a hyperbolic line ℓ in $\mathbb{H}$ so that n acts as reflection in ℓ; or

2 n fixes no point of $\mathbb{H}$, in which case n fixes exactly two points of $\overline{\mathbb{R}}$ and acts as a glide reflection along the hyperbolic line ℓ determined by these two points.

Exercise 6.5

Let $p(z) = z + 1$ be parabolic and let n be reflection in a hyperbolic line ℓ. Compute the composition $p \circ n$ and determine the fixed points of $p \circ n$.

6.2 Discreteness

Over the course of the next four sections, we give a very brief and sketchy introduction to what it means for a subgroup of $\text{Möb}(\mathbb{H})$ to be well-behaved. These sections should be viewed as the barest of teasers, and the interested reader should consult the list of Further Reading for a list of sources which treat these topics more in depth.

We begin by describing two notions of well-behavedness for a subgroup Γ of $\text{Möb}(\mathbb{H})$. We go on to show that they are in fact two manifestations of the same basic principle. As has been the case throughout the book, the fact that we choose to work in the upper half-plane model $\mathbb{H}$ is not essential, and all the results translate to the other models of the hyperbolic plane.

We begin with a topological definition.

Definition 6.4

A subset Z of $\mathbb{H}$ is *discrete* if for each $z \in Z$ there exists some $\varepsilon > 0$ so that $U_\varepsilon(z) \cap Z = \{z\}$, where

$$U_\varepsilon(z) = \{w \in \mathbb{H} \mid \mathrm{d}_{\mathbb{H}}(z, w) < \varepsilon\}$$

is the open hyperbolic disc with hyperbolic centre z and hyperbolic radius ε.

In words, a subset Z of $\mathbb{H}$ is *discrete* if each point of Z can be isolated from all the other points of Z. This definition of discreteness is very similar to some of the ideas surrounding closed sets and convergence explored in Section 1.2.

As a simple example, the set $X = \{\frac{1}{n} + i \,:\, n \in \mathbb{N}\}$ is a discrete subset of $\mathbb{H}$, since each natural number n we may take

$$\varepsilon = \max\left(\mathrm{d}_{\mathbb{H}}\left(\frac{1}{n} + i, \frac{1}{n-1} + i\right), \mathrm{d}_{\mathbb{H}}\left(\frac{1}{n} + i, \frac{1}{n+1} + i\right)\right).$$

However, the set $Y = \{i\} \cup \{\frac{1}{n} + i \,:\, n \in \mathbb{N}\}$ is not a discrete subset of $\mathbb{H}$. To see this, note that $\mathrm{d}_{\mathbb{H}}(i, \frac{1}{n} + i) < \frac{1}{n}$, and so $U_\varepsilon(i)$ contains $\frac{1}{n} + i$ for every n satisfying $n > \frac{1}{\varepsilon}$.

Exercise 6.6

Prove that a discrete subset of $\mathbb{H}$ is countable.

We may now use this definition of discrete sets in $\mathbb{H}$ to give a definition of what it means for a subgroup of $\text{Möb}(\mathbb{H})$ to be discrete.

Definition 6.5

A subgroup Γ of Möb($\mathbb{H}$) is *discrete* if the set

$$\Gamma(z) = \{\gamma(z) \mid \gamma \in \Gamma\}$$

is discrete for every $z \in \mathbb{H}$.

We refer to the set $\Gamma(z) = \{\gamma(z) \mid \gamma \in \Gamma\}$ as the *orbit* of z under Γ.

One difficulty with this definition is that it can be difficult to use it to prove that a given subgroup of Möb($\mathbb{H}$) is discrete. It is, on the other hand, sometimes very easy to use it to prove that a subgroup of Möb($\mathbb{H}$) is not discrete.

For example, we know by Exercise 2.27 that an elliptic element m_θ of Möb$^+$($\mathbb{H}$) that fixes i has the form

$$m_\theta(z) = \frac{\cos(\theta)z + \sin(\theta)}{-\sin(\theta)z + \cos(\theta)}$$

for some real number θ.

So, consider the subgroup

$$\Theta = \{m_\theta \mid \theta \in \mathbb{R}\}$$

of Möb$^+$($\mathbb{H}$).

Since every element of Θ fixes i, we have that $\Theta(i) = \{i\}$.

However, for any point $z_0 \neq i$ in $\mathbb{H}$, we have that

$$\Theta(z_0) = \{\gamma(z_0) \mid \gamma \in \Theta\} = \{w \in \mathbb{H} \mid d_{\mathbb{H}}(i, w) = d_{\mathbb{H}}(i, z_0)\},$$

which is the hyperbolic circle with hyperbolic centre i and hyperbolic radius $d_{\mathbb{H}}(i, z_0)$.

In particular, $\Theta(z_0)$ is not a discrete subset of $\mathbb{H}$ for $z_0 \neq i$, and so Θ is not a discrete subgroup of Möb($\mathbb{H}$). This example also shows why it is not enough to check the translates of a single point of $\mathbb{H}$, but instead why we must check every point of $\mathbb{H}$.

In general, given a subgroup Γ of Möb($\mathbb{H}$) and a point z of $\mathbb{H}$, consider the subgroup

$$\Gamma_z = \{\gamma \in \Gamma \mid \gamma(z) = z\}$$

of Γ consisting of the elements of Γ fixing z. We refer to Γ_z as the *stabilizer* of z in Γ.

For the subgroup $\Theta = \{m_\theta(z) \mid \theta \in \mathbb{R}\}$ considered above, the stabilizer Θ_i of i is all of Θ, while the stabilizer of any point $z \neq i$ is just the trivial group.

If we examine this group Θ a bit more carefully, we see that no infinite subgroup of Θ is a discrete subgroup of $\text{Möb}(\mathbb{H})$. To see this, let Θ^0 be an infinite subgroup of Θ and choose a point $z_0 \in \mathbb{H} - \{i\}$.

By Theorem 6.1, we have that no non-trivial element of $\text{Möb}^+(\mathbb{H})$ can fix two different points of $\mathbb{H}$. Hence, if m_{θ_1} and m_{θ_2} are distinct elements of Θ, then $m_{\theta_1}(z_0) \neq m_{\theta_2}(z_0)$.

In particular, since Θ^0 is an infinite subgroup of Θ, we have that $\Theta^0(z_0)$ is an infinite subset of the hyperbolic circle C with hyperbolic centre z_0 and hyperbolic radius $d_{\mathbb{H}}(z_0, i)$. Since C is compact, there then exists a sequence $\{m_{\theta_n}\}$ of elements of Θ^0 so that $\{m_{\theta_n}(z_0)\}$ converges to some point w of C.

Since $\{m_{\theta_n}(z_0)\}$ converges to w as $n \to \infty$, we have that $\{m_{\theta_{n+1}}^{-1} \circ m_{\theta_n}(z_0)\}$ converges to z_0 as $n \to \infty$. Since Θ^0 is a subgroup of $\text{Möb}(\mathbb{H})$, we have that each $m_{\theta_{n+1}}^{-1} \circ m_{\theta_n}$ is again an element of Θ^0. In particular, we have that the set $\Theta^0(z_0)$ is not a discrete subset of C, and so Θ^0 is not a discrete subgroup of $\text{Möb}(\mathbb{H})$.

This leads us to the following general result.

Theorem 6.6

Let Γ be a subgroup of $\text{Möb}(\mathbb{H})$. If Γ is discrete, then the stabilizer

$$\Gamma_x = \{\gamma \in \Gamma \mid \gamma(x) = x\}$$

is finite for every $x \in \mathbb{H}$.

We prove Theorem 6.6 by contradiction. Suppose there exists a point x of $\mathbb{H}$ so that Γ_x is infinite. The argument that for any $w \neq x$ we have that $\Gamma_x(w)$ is a non-discrete subset of the hyperbolic circle with hyperbolic centre x and hyperbolic radius $d_{\mathbb{H}}(x, w)$ is exactly the same as the argument given above that $\Theta^0(z_0)$ is a non-discrete subset of the hyperbolic circle with hyperbolic centre i and hyperbolic radius $d_{\mathbb{H}}(z_0, i)$. This completes the proof of Theorem 6.6.

In general, the converse to Theorem 6.6 does not hold. For instance, consider the subgroup

$$\Phi = \{m_\lambda(z) = \lambda z \mid \lambda > 0\}$$

of $\text{Möb}(\mathbb{H})$. This group Φ is not a discrete subgroup of $\text{Möb}(\mathbb{H})$, since $\Phi(i)$ is equal to the positive imaginary axis. However, the stabilizer Φ_z is trivial for every z in $\mathbb{H}$.

Exercise 6.7

Prove that every discrete subgroup of

$$\Gamma = \{m_\lambda(z) = \lambda z \mid \lambda > 0\}$$

is infinite cyclic.

Exercise 6.8

Prove that a discrete subgroup of Möb($\mathbb{H}$) is countable.

There is an useful fact about discrete groups, which has so far come up twice in similar guises, in the argument preceding the statement and proof of Theorem 6.6 and in the solution to Exercise 6.8. We record this fact in the following proposition.

Proposition 6.7

If Γ is a discrete subgroup of Möb($\mathbb{H}$) and if $z \in \mathbb{H}$ is any point, then $\Gamma(z) \cap U_\varepsilon(z)$ is finite for every $\varepsilon > 0$.

To prove Proposition 6.7, we argue by contradiction. Suppose there exists some point z in $\mathbb{H}$ and some $\varepsilon > 0$ so that $\Gamma(z) \cap U_\varepsilon(z)$ is infinite. Since $\overline{U_\varepsilon(z)}$ is compact, there exists a sequence $\{\gamma_n\}$ in Γ and a point w in $\overline{U_\varepsilon(z)}$ so that $\{\gamma_n(z)\}$ converges to w.

Since $\{\gamma_n(z)\}$ converges to w, we have that $\{\gamma_{n+1}^{-1} \circ \gamma_n(z)\}$ converges to z. Since the identity belongs to Γ, z is contained in $\Gamma(z)$. Since Γ is a group, each $\gamma_{n+1}^{-1} \circ \gamma_n$ is itself an element of Γ, and so each $\gamma_{n+1}^{-1} \circ \gamma_n(z)$ is contained in $\Gamma(z)$. Hence, we have that $\Gamma(z)$ is not discrete. This contradiction completes the proof of Proposition 6.7.

There is another common way of defining discreteness, if we restrict our attention to subgroups of Möb$^+$($\mathbb{H}$). Let Γ be a subgroup of Möb$^+$($\mathbb{H}$), and suppose that Γ is not discrete. That is, there is some $z \in \mathbb{H}$ so that the set $\Gamma(z)$ is not a discrete subset of $\mathbb{H}$.

By the definition of discreteness, there then exists an element $\gamma(z)$ of $\Gamma(z)$ so that for each $\varepsilon > 0$, the set $\Gamma(z) \cap U_\varepsilon(\gamma(z))$ contains a point other than $\gamma(z)$. For each $n \in \mathbb{N}$, choose an element γ_n of Γ so that $\gamma_n(z) \neq \gamma(z)$ and so that

$$\gamma_n(z) \in \Gamma(z) \cap U_{\frac{1}{n}}(\gamma(z)).$$

As $n \to \infty$, we have that $d_{\mathbb{H}}(\gamma(z), \gamma_n(z)) \to 0$. Pass to a subsequence of $\{\gamma_n\}$, again called $\{\gamma_n\}$ to avoid the proliferation of subscripts, so that the $\gamma_n(z)$ are distinct. We now have a sequence $\{\gamma_n\}$ of distinct elements of Γ so that $\{\gamma_n(z)\}$ converges to $\gamma(z)$.

Write $\gamma_n(z) = \frac{a_n z + b_n}{c_n z + d_n}$ and $\gamma(z) = \frac{az+b}{cz+d}$, all normalized to have determinant equal to 1. One way to force $\{\gamma_n(z)\}$ to converge to $\gamma(z)$ is to have $a_n \to a$, $b_n \to b$, $c_n \to c$, and $d_n \to d$ as $n \to \infty$. Note that the converse of this argument holds true as well.

Definition 6.8

A sequence $\{\gamma_n(z) = \frac{a_n z + b_n}{c_n z + d_n}\}$ of elements of $\text{Möb}^+(\mathbb{H})$ *converges* to the element $\gamma(z) = \frac{az+b}{cz+d}$ of $\text{Möb}^+(\mathbb{H})$ if $a_n \to a$, $b_n \to b$, $c_n \to c$, and $d_n \to d$ as $n \to \infty$.

Exercise 6.9

Let Γ be a subgroup of $\text{Möb}^+(\mathbb{H})$. Prove that Γ contains a sequence of distinct elements converging to an element φ of $\text{Möb}^+(\mathbb{H})$ if and only if Γ contains a sequence of distinct elements converging to the identity.

This formulation of the definition of discreteness in $\text{Möb}^+(\mathbb{H})$ immediately yields some specific examples of discrete subgroups of $\text{Möb}^+(\mathbb{H})$. For instance, the *modular group*

$$\text{PSL}_2(\mathbb{Z}) = \left\{ m(z) = \frac{az+b}{cz+d} \mid a,\ b,\ c,\ d \in \mathbb{Z} \text{ and } ad - bc = 1 \right\}$$

is discrete, as are all of its subgroups.

In general, a discrete subgroup of $\text{Möb}^+(\mathbb{H})$ is called a *Fuchsian group*.

In general, if Φ is a subgroup of Γ, then $\Phi(z)$ is a subset of $\Gamma(z)$ for every point z of $\mathbb{H}$, and so subgroups of discrete groups are themselves discrete. We record this as Proposition 6.9.

Proposition 6.9

Let Γ be a discrete subgroup of $\text{Möb}(\mathbb{H})$. If Φ is a subgroup of Γ, then Φ is discrete.

Conversely, there are a few special cases in which the discreteness of a subgroup of Γ implies the discreteness of Γ. We begin by considering subgroups of $\text{Möb}^+(\mathbb{H})$ with discrete normal subgroups.

Proposition 6.10

Let Γ be a non-trivial subgroup of $\text{Möb}^+(\mathbb{H})$ and let Φ be a non-trivial normal subgroup of Γ. If Φ is discrete, then Γ is discrete.

To prove Proposition 6.10, we use the contrapositive. Suppose that Γ is not discrete, and let $\{\gamma_n\}$ be a sequence of distinct elements of Γ converging to the identity.

Choose some element φ of Φ, other than the identity, and consider the sequence $\{\gamma_n^{-1} \circ \varphi \circ \gamma_n\}$. Since Φ is a normal subgroup of Γ, each $\gamma_n^{-1} \circ \varphi \circ \gamma_n$ is contained in Φ, and so $\{\gamma_n^{-1} \circ \varphi \circ \gamma_n\}$ is a sequence of elements of Φ.

Since $\{\gamma_n\}$ converges to the identity, we have that $\{\gamma_n^{-1}\}$ converges to the identity as well, and so $\{\gamma_n^{-1} \circ \varphi \circ \gamma_n\}$ converges to φ. All that remains to show is that the $\gamma_n^{-1} \circ \varphi \circ \gamma_n$ are distinct, which follows immediately from the facts that the γ_n are distinct and are converging to the identity. This completes the proof of Proposition 6.10.

There is also the following result about subgroups of $\text{Möb}^+(\mathbb{H})$ containing a discrete subgroup of finite index.

Proposition 6.11

Let Γ be a subgroup of $\text{Möb}(\mathbb{H})$, and let Φ be a finite index subgroup of Γ. If Φ is discrete, then Γ is discrete.

We begin the proof of Proposition 6.11 by expressing Γ as a coset decomposition with respect to the subgroup Φ. That is, we write

$$\Gamma = \cup_{k=0}^{p} \alpha_k \, \Phi,$$

where $\alpha_0, \ldots, \alpha_p$ are elements of Γ.

Suppose that Γ is not discrete, and let $\{\gamma_n\}$ be a sequence of distinct elements of Γ converging to the identity. For each n, we can write $\gamma_n = \alpha_{k_n} \varphi_n$, where $0 \le k_n \le p$ and $\varphi_n \in \Phi$.

Since there are infinitely many elements in the sequence, there is some fixed q satisfying $0 \le q \le p$, so that $k_n = q$ for infinitely many n. So, consider the subsequence $\{\gamma_m = \alpha_q \varphi_m\}$ consisting of those elements of the sequence for which $k_n = q$.

Since $\{\gamma_m\}$ converges to the identity, we have that $\{\alpha_q \varphi_m\}$ converges to the identity as well. In particular, we have that $\{\varphi_m\}$ converges to α_q^{-1}. By Exercise

6.9, this implies Φ is not discrete, a contradiction. This completes the proof of Proposition 6.11.

Exercise 6.10

Two subgroups Φ and Θ of a subgroup Γ of $\text{Möb}^+(\mathbb{H})$ are *commensurable* if $\Phi \cap \Theta$ has finite index in both Φ and Θ. Prove that if Φ and Θ are commensurable subgroups of Γ, then Φ is discrete if and only if Θ is discrete.

6.3 Fundamental Polygons

One method of investigating the question of constructing discrete subgroups of $\text{Möb}(\mathbb{H})$ is to first compile a list of properties that a discrete subgroup of $\text{Möb}(\mathbb{H})$ has. We can then examine these properties, and attempt to prove that a subgroup of $\text{Möb}(\mathbb{H})$ that satisfies all the properties on the list is necessarily discrete.

So, let Γ be a discrete subgroup of $\text{Möb}(\mathbb{H})$. We know from Exercise 6.8 that Γ is necessarily countable. We also know from Theorems 6.1 and 6.3 that the fixed point set of every element of Γ is either empty, a point in $\mathbb{H}$, or a hyperbolic line in $\mathbb{H}$.

In particular, since we cannot express $\mathbb{H}$ as a countable union of points and hyperbolic lines, most points of $\mathbb{H}$ are not fixed by any non-trivial element of Γ. For those readers who are interested, the proof of this statement is an immediate application of the Baire category theorem. We refer the interested reader to Munkres [19] for more information on this topic.

We also note that we are being deliberately vague about what is meant by the phrase 'most points of $\mathbb{H}$'. For the purposes of this section and the next, it suffices that there exists one point of $\mathbb{H}$ not fixed by any non-trivial element of Γ.

In this section, we use this fact that there exists a point of $\mathbb{H}$ that is not fixed by any non-trivial element of Γ to construct a hyperbolic polygon that encodes the action of Γ on $\mathbb{H}$. We need to make a few definitions to start.

The action of the discrete subgroup Γ of $\text{Möb}(\mathbb{H})$ on $\mathbb{H}$ induces an equivalence relation $\sim_\Gamma$ on $\mathbb{H}$, where two points z and w of $\mathbb{H}$ are equivalent, denoted $z \sim_\Gamma w$, if there exists an element γ of Γ with $w = \gamma(z)$.

The fact that $\sim_\Gamma$ is an equivalence relation follows immediately from the fact that Γ is a group acting on $\mathbb{H}$. Namely, for each $z \in \mathbb{H}$ we have that $z \sim_\Gamma z$ since the identity is an element of Γ. Also, if we have that $z \sim_\Gamma w$, so that there is an element γ of Γ with $w = \gamma(z)$, then $z = \gamma^{-1}(w)$ and so $w \sim_\Gamma z$.

To check transitivity, assume that $z \sim_\Gamma w$, so that there exists an element γ of Γ with $w = \gamma(z)$, and that $w \sim_\Gamma y$, so that there exists an element φ of Γ with $y = \varphi(w)$. Since

$$y = \varphi(w) = \varphi(\gamma(z)) = (\varphi \circ \gamma)(z)$$

and since $\varphi \circ \gamma$ is an element of Γ, we have that $y \sim_\Gamma z$, and so $\sim_\Gamma$ is transitive.

The equivalence classes of $\sim_\Gamma$ are exactly the orbits $\Gamma(z)$ of points z in $\mathbb{H}$ under Γ, as defined and discussed in Section 6.2.

Definition 6.12

A *fundamental set* for the action of Γ on $\mathbb{H}$ is the choice of one point from each equivalence class determined by the equivalence relation $\sim_\Gamma$.

Fundamental sets can be extremely badly behaved, as we are imposing no conditions on how we are choosing points from the equivalence classes of $\sim_\Gamma$.

As a theoretical construction, fundamental sets are very useful, and indeed are inescapable. However, as a practical construction, they are too flexible to be effective. For discrete subgroups of Möb($\mathbb{H}$), we can refine the notion of a fundamental set to obtain a more useful object. We begin by describing a *fundamental domain.*

We begin with a pair of definitions. One is a generalization of the stabilizer of a point discussed in Section 6.2.

Definition 6.13

Given a discrete subgroup Γ of Möb($\mathbb{H}$) and a subset X of $\mathbb{H}$, define the *stabilizer* $\text{stab}_\Gamma(X)$ of X in Γ to be the subgroup

$$\text{stab}_\Gamma(X) = \{\gamma \in \Gamma \mid \gamma(X) = X\}$$

of Γ.

In the case that X consists of a single point $X = \{x\}$, the two definitions of the stabilizer of x in Γ coincide. That is, we have that

$$\Gamma_x = \text{stab}_\Gamma(\{x\}).$$

To take another example, let Γ be a discrete subgroup of Möb($\mathbb{H}$), and let ℓ be a hyperbolic line in $\mathbb{H}$. There are several possibilities for the stabilizer $\text{stab}_\Gamma(\ell)$ of ℓ in Γ. One is that $\text{stab}_\Gamma(\ell)$ is trivial.

Suppose now that $\text{stab}_\Gamma(\ell)$ is non-trivial, so that there exists a non-trivial element of Γ taking ℓ to itself. In this case, we use Theorem 6.1 and Theorem 6.3 to determine $\text{stab}_\Gamma(\ell)$.

The most general case is that there is both an element of Γ that is reflection in ℓ and a loxodromic element of Γ whose axis is ℓ. In this case, we have that $\text{stab}_\Gamma(\ell) = \mathbb{Z} \oplus \mathbb{Z}_2$, where $\mathbb{Z}_2$ is the quotient group $\mathbb{Z}_2 = \mathbb{Z}/2\mathbb{Z}$.

If there is no element of Γ that is reflection in ℓ, then either there is a loxodromic element of Γ whose axis is ℓ, or there is a glide reflection in Γ that takes ℓ to itself. In both of these cases, $\text{stab}_\Gamma(\ell)$ is infinite cyclic.

The only remaining case is that there is an element of Γ that is reflection in ℓ, and no other non-trivial element of Γ taking ℓ to itself. In this case, we have that $\text{stab}_\Gamma(\ell) = \mathbb{Z}_2$.

There is a second definition we need.

Definition 6.14

Let Γ be a discrete subgroup of Möb($\mathbb{H}$), and let Φ be a subgroup of Γ. A set X in $\mathbb{H}$ is *precisely invariant under Φ in Γ* if two conditions are met, namely that $\Phi = \text{stab}_\Gamma(X)$ and that if γ is an element of Γ for which $X \cap \gamma(X) \neq \emptyset$, then γ is actually an element of Φ.

As an example, consider the case in which X is a hyperbolic line in $\mathbb{H}$ and $\Phi = \text{stab}_\Gamma(X)$. In this case, the first of the two conditions is met by the definition of Φ.

The second condition is met if and only for each pair γ_1 and γ_2 of elements of Γ, we have either that $\gamma_1(X) = \gamma_2(X)$, in which case $\gamma_2^{-1} \circ \gamma_1$ is an element of $\text{stab}_\Gamma(X)$, or that $\gamma_1(X)$ and $\gamma_2(X)$ are disjoint.

In the special case that Φ is the trivial subgroup containing only the identity, we say that X is *precisely invariant under the identity* in Γ. Rephrasing this, we have that X is precisely invariant under the identity in Γ if and only if $X \cap \gamma(X) = \emptyset$ for all non-trivial elements γ of Γ.

As an example, let D be the open region in $\mathbb{H}$ bounded by the hyperbolic line ℓ_1 determined by -1 and 1, and the hyperbolic line ℓ_2 determined by -2 and 2. For each $\lambda > 1$, let $m_\lambda(z) = \lambda z$, and consider the subgroup

$$\Gamma_\lambda = \langle m_\lambda \rangle = \{ m_\lambda^k(z) = \lambda^k z \mid k \in \mathbb{Z} \}$$

of Möb($\mathbb{H}$) generated by m_λ. We can determine exactly for which λ the set D is precisely invariant under the identity in Γ_λ.

Take $\lambda \geq 2$. For $k > 0$, both of the hyperbolic lines $m_\lambda^k(\ell_1)$ and $m_\lambda^k(\ell_2)$ lie in the closed half-plane determined by ℓ_2 that does not contain D. In particular, $m_\lambda^k(D)$ is disjoint from D.

Similarly, for $k < 0$, both of the hyperbolic lines $m_\lambda^k(\ell_1)$ and $m_\lambda^k(\ell_2)$ lie in the closed half-plane determined by ℓ_1 that does not contain D. In particular, $m_\lambda^k(D)$ is disjoint from D.

Hence, $D \cap m_\lambda^k(D) \neq \emptyset$ only for $k = 0$, in which case m_λ^k is the identity and so $m_\lambda^k(D) = D$. So, D is precisely invariant under the identity in Γ_λ for $\lambda \geq 2$.

However, for $1 < \lambda < 2$, the hyperbolic line $m_\lambda(\ell_1)$ lies inside D. Hence, $D \cap m_\lambda(D) \neq \emptyset$ and $D \cap m_\lambda(D) \neq D$. Hence, D is not precisely invariant under the identity for $1 < \lambda < 2$.

Definition 6.15

Let Γ be a discrete subgroup of Möb($\mathbb{H}$). A *fundamental domain* for the action of Γ on $\mathbb{H}$ is an open set U in $\mathbb{H}$ that satisfies two conditions, namely that U is precisely invariant under the identity in Γ and that its closure $\overline{U}$ in $\mathbb{H}$ contains a fundamental set for the action of Γ.

Exercise 6.11

Prove that for a discrete subgroup Γ of Möb($\mathbb{H}$) and a set U in $\mathbb{H}$, the closure $\overline{U}$ in $\mathbb{H}$ contains a fundamental set for the action of Γ if and only if the union

$$\Gamma(\overline{U}) = \cup_{\gamma \in \Gamma} \gamma(\overline{U})$$

is equal to $\mathbb{H}$.

For the set D considered above, bounded by the hyperbolic lines ℓ_1 and ℓ_2, we have already shown that D is precisely invariant under the identity in Γ_λ for all $\lambda \geq 2$.

If we take $\lambda = 2$, we see that $m_2(\ell_1) = \ell_2$, and so the union of the translates of $\overline{D}$ by elements of Γ_2 is equal to all of $\mathbb{H}$. Applying Exercise 6.11 we see that $\overline{D}$ contains a fundamental set for the action of Γ_2 on $\mathbb{H}$. Since D is precisely invariant under the identity in Γ_2, this shows that D is a fundamental domain for Γ_2.

On the other hand, for every $\lambda > 2$, there is no point in the closure $\overline{D}$ of D that is equivalent under Γ_λ to $\frac{1}{2}(2 + \lambda)\, i$. Since the union of the translates of

$\overline{D}$ is not equal to $\mathbb{H}$, $\overline{D}$ does not contain a fundamental set for the action of Γ_λ on $\mathbb{H}$. Hence, D is not a fundamental domain for Γ_λ.

As in this example, it can sometimes happen that the closure of a fundamental domain is a hyperbolic polygon, though it should be noted that it is not part of the definition that the closure of a fundamental domain be a hyperbolic polygon.

For example, the open set F in $\mathbb{H}$ bounded by the Euclidean lines

$$\{z \in \mathbb{H} \mid \mathrm{Im}(z) = 1\} \text{ and } \{z \in \mathbb{H} \mid \mathrm{Im}(z) = 2\}$$

is also a fundamental domain for Γ_2. However, the closure $\overline{F}$ of F is not a hyperbolic polygon, since $\overline{F}$ is not convex.

So, we can refine the notion of a fundamental domain one step further.

Definition 6.16

A *fundamental polygon* for a discrete subgroup Γ of Möb($\mathbb{H}$) is a hyperbolic polygon P whose interior is a fundamental domain for Γ.

As an example, the closure $\overline{D}$ of the fundamental domain D for the group Γ_λ with $\lambda = 2$ discussed above is a fundamental polygon. In fact, we can express $\overline{D}$ as the intersection of a pair of closed half-planes in $\mathbb{H}$.

6.4 The Dirichlet Polygon

There is a general construction for fundamental polygons that works for any discrete subgroup Γ of Möb($\mathbb{H}$). Let z_0 be any point of $\mathbb{H}$ that is not fixed by any non-trivial element of Γ. We saw at the beginning of Section 6.3 that such a point necessarily exists.

Let Γ' be the subset of all non-trivial elements of Γ. For each element γ of Γ', consider the hyperbolic line ℓ_γ that is the perpendicular bisector of the hyperbolic line segment between z_0 and $\gamma(z_0)$. Let H_γ be the closed half-plane determined by γ that contains z_0. Consider the intersection

$$D_\Gamma(z_0) = \cap_{\gamma \in \Gamma'} H_\gamma.$$

This is the *Dirichlet polygon* for Γ centred at z_0. The main result of this section is that the Dirichlet polygon is a fundamental polygon for Γ.

Theorem 6.17

Let Γ be a discrete subgroup of $\text{Möb}(\mathbb{H})$ and let z_0 be a point of $\mathbb{H}$ not fixed by any non-trivial element of Γ. Then, the Dirichlet polygon $D_\Gamma(z_0)$ for Γ centred at z_0 is a fundamental polygon for Γ.

There are two things to show: that $D_\Gamma(z_0)$ is a hyperbolic polygon and that $D_\Gamma(z_0)$ is a fundamental polygon for Γ.

Since $D_\Gamma(z_0)$ is the intersection of the collection $\mathcal{H} = \{H_\gamma \mid \gamma \in \Gamma'\}$ of closed half-planes in $\mathbb{H}$, we have that $D_\Gamma(z_0)$ is a closed convex subset of $\mathbb{H}$. In order to complete the proof that $D_\Gamma(z_0)$ is a hyperbolic polygon, we need only show that $\mathcal{H}$ is locally finite.

The local finiteness of $\mathcal{H}$ follows immediately from Proposition 6.7. Given $\varepsilon > 0$, we know by the construction of ℓ_γ that $\ell_\gamma \cap U_\varepsilon(z_0) \neq \emptyset$ if and only if $\gamma(z_0) \in U_{2\varepsilon}(z_0)$. Since the intersection $\Gamma(z_0) \cap U_{2\varepsilon}(z_0)$ contains only finitely many translates of z_0, by Proposition 6.7, we have that only finitely many of the ℓ_γ intersect $U_\varepsilon(z_0)$.

For any point w of $\mathbb{H}$ and any $C > 0$, we have that $U_C(w)$ is contained in $U_\varepsilon(z_0)$ for some $\varepsilon > 0$, namely $\varepsilon = d_{\mathbb{H}}(z_0, w) + C$, and so only finitely many of the ℓ_γ intersect $U_C(w)$. This completes the proof of the claim that $\mathcal{H}$ is locally finite.

We begin the proof that $D_\Gamma(z_0)$ is a fundamental polygon for Γ by giving a slightly different but equivalent definition of $D_\Gamma(z_0)$. First, recall that the perpendicular bisector ℓ_γ of the hyperbolic line segment joining z_0 and $\gamma(z_0)$ is characterized by the property that

$$\ell_\gamma = \{w \in \mathbb{H} \mid d_{\mathbb{H}}(z_0, w) = d_{\mathbb{H}}(\gamma(z_0), w)\}.$$

In particular, the closed half-plane H_γ determined by ℓ_γ is the set of points in $\mathbb{H}$ closer to z_0 than to $\gamma(z_0)$. That is,

$$H_\gamma = \{w \in \mathbb{H} \mid d_{\mathbb{H}}(z_0, w) \leq d_{\mathbb{H}}(\gamma(z_0), w)\}.$$

So, we can also describe $D_\Gamma(z_0)$ as

$$D_\Gamma(z_0) = \{w \in \mathbb{H} \mid d_{\mathbb{H}}(z_0, w) \leq d_{\mathbb{H}}(\gamma(z_0), w) \text{ for all } \gamma \in \Gamma'\}.$$

In particular, the interior $\text{int}(D_\Gamma(z_0))$ of $D_\Gamma(z_0)$ is the set

$$\text{int}(D_\Gamma(z_0)) = \{w \in \mathbb{H} \mid d_{\mathbb{H}}(z_0, w) < d_{\mathbb{H}}(\gamma(z_0), w) \text{ for all } \gamma \in \Gamma'\}.$$

In words, $\text{int}(D_\Gamma(z_0))$ is the set of all points of $\mathbb{H}$ that are closer to z_0 than to any other point of $\Gamma(z_0)$.

We first show that $\text{int}(D_\Gamma(z_0))$ is precisely invariant under the identity in Γ. If not, then there exists a non-trivial element φ of Γ so that the intersection

$$\text{int}(D_\Gamma(z_0)) \cap \varphi(\text{int}(D_\Gamma(z_0)))$$

contains a point x.

Since x is a point of $\text{int}(D_\Gamma(z_0))$, we have that

$$\text{d}_{\mathbb{H}}(x, z_0) < \text{d}_{\mathbb{H}}(x, \gamma(z_0)) \text{ for all } \gamma \in \Gamma'.$$

Since x is a point of $\varphi(\text{int}(D_\Gamma(z_0)))$, we can write $x = \varphi(y)$, where y is a point of $\text{int}(D_\Gamma(z_0))$. Since y is a point of $\text{int}(D_\Gamma(z_0))$, we have that

$$\text{d}_{\mathbb{H}}(y, z_0) < \text{d}_{\mathbb{H}}(y, \varphi^{-1}(z_0)).$$

Since φ is an isometry of $\mathbb{H}$, we have that

$$\text{d}_{\mathbb{H}}(x, \varphi(z_0)) = \text{d}_{\mathbb{H}}(\varphi^{-1}(x), z_0) = \text{d}_{\mathbb{H}}(y, z_0)$$

and that

$$\text{d}_{\mathbb{H}}(y, \varphi^{-1}(z_0)) = \text{d}_{\mathbb{H}}(\varphi(y), z_0) = \text{d}_{\mathbb{H}}(x, z_0).$$

Combining these, we have that

$$\text{d}_{\mathbb{H}}(x, \varphi(z_0)) < \text{d}_{\mathbb{H}}(x, z_0).$$

This, however, contradicts that x is a point of $\text{int}(D_\Gamma(z_0))$.

This shows that $\text{int}(D_\Gamma(z_0)))$ is precisely invariant under the identity in Γ. To complete the proof of Theorem 6.17, we need to know that $D_\Gamma(z_0)$ contains a fundamental set for the action of Γ on $\mathbb{H}$. So, let w be any point of $\mathbb{H}$.

By Proposition 6.7, there is some point in the orbit $\Gamma(z_0) = \{\gamma(z_0) \mid \gamma \in \Gamma\}$ that is closest to w. To see this, choose δ so that $U_\delta(w) \cap \Gamma(z_0)$ is non-empty, and let $\varepsilon = \delta + \text{d}_{\mathbb{H}}(z_0, w)$. Since $U_\delta(w) \subset U_\varepsilon(z_0)$ and since $U_\varepsilon(z_0) \cap \Gamma(z_0)$ is finite, we have that $U_\delta(w) \cap \Gamma(z_0)$ is finite. We may now choose a point $\gamma_w(z_0)$ in $U_\delta(w) \cap \Gamma(z_0)$ that is closest to w, though we should note that there is in general no way to guarantee that this point is unique.

If we now write out what it means for $\gamma_w(z_0)$ to be closest to w, we get

$$\text{d}_{\mathbb{H}}(\gamma_w(z_0), w) \leq \text{d}_{\mathbb{H}}(\gamma(z_0), w) \text{ for all } \gamma \in \Gamma.$$

Since γ_w is an isometry of $\mathbb{H}$, we can rewrite this as

$$\text{d}_{\mathbb{H}}(z_0, \gamma_w^{-1}(w)) \leq \text{d}_{\mathbb{H}}(\gamma_w^{-1} \circ \gamma(z_0), \gamma_w^{-1}(w)) \text{ for all } \gamma \in \Gamma.$$

Since $\gamma_w^{-1} \circ \gamma$ ranges over all of Γ as γ ranges over all of Γ, this implies that $\gamma_w^{-1}(w)$ lies in $D_\Gamma(z_0)$, by the definition of $D_\Gamma(z_0)$.

Since $D_\Gamma(z_0)$ contains a point in the orbit $\Gamma(w)$ of w for every point w of $\mathbb{H}$, we have that $D_\Gamma(z_0)$ contains a fundamental set for the action of Γ on $\mathbb{H}$. This completes the proof of Theorem 6.17.

In practice, calculating the Dirichlet polygon for a given discrete subgroup Γ of $\text{Möb}(\mathbb{H})$, centred at a particular point z_0 of $\mathbb{H}$, can be very difficult. Among other reasons, it may be very difficult to determine the orbit $\Gamma(z_0)$ of z_0, and so to determine the points of $\mathbb{H}$ closer to z_0 than to any other point in the orbit.

Also, we do not yet have many examples of discrete subgroups of $\text{Möb}(\mathbb{H})$ for which to calculate Dirichlet polygons. There is the modular group $\text{PSL}_2(\mathbb{Z})$ and its subgroups. In order to get a feel for how Dirichlet polygons behave, we work with a simpler example.

Fix a complex number τ with $\text{Im}(\tau) > 0$, and consider the group

$$\Theta = \{\theta_{n+m\tau}(z) = z + n + m\tau \mid n,\, m \in \mathbb{Z}\}$$

of homeomorphisms of $\mathbb{C}$.

Since the orbit $\Theta(z_0) = \{z_0 + n + m\tau \mid n,\, m \in \mathbb{Z}\}$ is a discrete subset of $\mathbb{C}$ for each $z_0 \in \mathbb{C}$, Θ is a discrete group of homeomorphisms of $\mathbb{C}$. We may now use the same definitions and arguments in $\mathbb{C}$ as we used in $\mathbb{H}$ to define the Dirichlet polygon $D_\Theta(z_0)$ for Θ centred at $z_0 \in \mathbb{C}$.

For example, take $\tau = i$, so that

$$\Theta = \{\theta_{n+mi}(z) = z + n + mi \mid n,\, m \in \mathbb{Z}\}.$$

The orbit of a point z_0 is then

$$\Theta(z_0) = \{z_0 + n + mi \mid n,\, m \in \mathbb{Z}\}.$$

Set $z_0 = 2$. To construct $D_\Theta(2)$, we need to find the perpendicular bisector $\ell_{m,n}$ of the Euclidean line segment joining 2 and $2 + n + mi$ for all $m,\, n \in \mathbb{Z}$.

For $n = 1$ and $m = 0$, $\ell_{1,0}$ is the vertical line $\{z \in \mathbb{H} \mid \text{Re}(z) = \frac{5}{2}\}$. For $n = -1$ and $m = 0$, $\ell_{-1,0}$ is the vertical line $\{z \in \mathbb{H} \mid \text{Re}(z) = \frac{3}{2}\}$.

For $n = 0$ and $m = 1$, $\ell_{0,1}$ is the horizontal line $\{z \in \mathbb{H} \mid \text{Im}(z) = \frac{1}{2}\}$. For $n = 0$ and $m = -1$, $\ell_{0,-1}$ is the horizontal line $\{z \in \mathbb{H} \mid \text{Im}(z) = -\frac{1}{2}\}$.

These four lines bound the square S in $\mathbb{C}$ with vertices at $\frac{3}{2} + \frac{1}{2}i$, $\frac{3}{2} - \frac{1}{2}i$, $\frac{5}{2} + \frac{1}{2}i$, and $\frac{5}{2} - \frac{1}{2}i$. The four perpendicular bisectors $\ell_{1,1}$, $\ell_{3,1}$, $\ell_{1,-1}$, and $\ell_{3,-1}$ pass through the vertices of S, and all the other perpendicular bisectors $\ell_{n,m}$ are disjoint from S.

Hence, in this case, the Dirichlet polygon $D_\Theta(2)$ is equal to the square S in $\mathbb{C}$ with vertices at $\frac{3}{2} + \frac{1}{2}i$, $\frac{3}{2} - \frac{1}{2}i$, $\frac{5}{2} + \frac{1}{2}i$, and $\frac{5}{2} - \frac{1}{2}i$.

Exercise 6.12

Draw $D_\Theta(z_0)$ for $\tau = 1 + 2i$ and $z_0 = 0$.

We close this section by noting that for these groups $\Theta = \{n + m\tau \mid n, m \in \mathbb{Z}\}$, the Dirichlet polygon $D_\Theta(z_0)$ need not be rectangular. It can also be hexagonal. For a more detailed discussion of these groups and their basic properties, we refer the interested reader to the book of Jones and Singerman [14].

Also, there are other constructions for fundamental polygons, such as the construction of the Ford domain. For further information, we refer the interested reader to the books of Maskit [18] and Ford [9].

6.5 Poincaré's Theorem

The construction from Section 6.4 of the Dirichlet polygon $D_\Gamma(z_0)$ gives a fundamental polygon for a discrete subgroup Γ of Möb($\mathbb{H}$). In practice, however, we are often more interested in using a hyperbolic polygon to construct a discrete subgroup Γ of Möb($\mathbb{H}$).

So, we begin this section by examining in more detail the construction of the Dirichlet polygon $D_\Gamma(z_0)$, in order to determine what collection of properties a hyperbolic polygon needs to satisfy in order to be used to construct a discrete subgroup of Möb($\mathbb{H}$).

The first observation is that we can associate a non-trivial element of Γ to each side of $D_\Gamma(z_0)$. To see this, let s be a side of $D_\Gamma(z_0)$ and let ℓ_s be the hyperbolic line containing s. By the definition of the Dirichlet polygon, there is a non-trivial element γ_s of Γ so that ℓ_s is the perpendicular bisector of the hyperbolic line segment joining z_0 and $\gamma_s(z_0)$. We associate this element γ_s of Γ to the side s of $D_\Gamma(z_0)$.

Note that, since γ_s is an isometry of $\mathbb{H}$ and since $D_\Gamma(z_0)$ is defined in terms of hyperbolic distance to z_0, the hyperbolic line $\gamma_s^{-1}(\ell_s)$ also contains a side s' of $D_\Gamma(z_0)$. Moreover, we have that $\gamma_s(s') = s$, that $\gamma_{s'} = \gamma_s^{-1}$, and that the intersection of $D_\Gamma(z_0)$ and its translate $\gamma_s(D_\Gamma(z_0))$ satisfies

$$\gamma_s(D_\Gamma(z_0)) \cap D_\Gamma(z_0) = s.$$

This analysis gives rise to the following definition.

Definition 6.18

Let P be a hyperbolic polygon. An element γ of Möb($\mathbb{H}$) is a *side pairing transformation* for P if there are two sides s and s' of P, which are not necessarily distinct, so that $\gamma(s') = s$, so that $\gamma_{s'} = \gamma_s^{-1}$, and so that $\gamma(P) \cap P = s$.

As a simple example, let T be a compact hyperbolic triangle in $\mathbb{H}$. Label the sides of T as s_1, s_2, and s_3. Let ℓ_k be the hyperbolic line in $\mathbb{H}$ containing s_k, and let γ_k denote the reflection in ℓ_k.

For each $1 \leq k \leq 3$, the element γ_k of Möb($\mathbb{H}$) is a side pairing transformation for P, since $\gamma_k(s_k) = s_k$, since $\gamma_k = \gamma_k^{-1}$, and since $\gamma_k(T) \cap T = s_k$.

Actually, this is just a single case of a more general construction. Take T to be any hyperbolic polygon, and label the sides of T as $s_1, \ldots, s_k, \ldots$. Let ℓ_k be the hyperbolic line containing s_k, and let γ_k denote reflection in ℓ_k. Then, each γ_k is a side pairing transformation of T, pairing s_k with itself.

Working in another model, let P_α be the compact regular octagon in the Poincaré disc $\mathbb{D}$ whose interior angles are all equal to α. By Proposition 5.17, we know that such a P_α exists for all α in the interval $(0, \frac{3}{4}\pi)$.

Label the sides of P_α counterclockwise as $s_1, \ldots, s_8$. For $1 \leq k \leq 4$, let γ_k be the loxodromic element of Möb$^+$($\mathbb{D}$) whose axis is the perpendicular bisector of both s_k and s_{4+k} and that satisfies $\gamma_k(s_{4+k}) = s_k$. Note that γ_k is a side pairing transformation for P_α, since $\gamma_k(s_{4+k}) = s_k$, since $\gamma_{s_{4+k}} = \gamma_{s_k}^{-1}$, and since $\gamma_k(P_\alpha) \cap P_\alpha = s_k$.

By virtue of its construction, every Dirichlet polygon comes equipped with a collection of side pairing transformations. We now wish to see what conditions the side pairing transformations of an arbitrary hyperbolic polygon P need to satisfy, in order to assure that the side pairing transformations generate a discrete subgroup Γ of Möb($\mathbb{H}$) with P as a fundamental polygon.

So, let P be a compact hyperbolic polygon and let $\mathcal{G}$ be a *complete collection of side pairing transformations* for P. By this, we mean that $\mathcal{G}$ is a collection of elements of Möb($\mathbb{H}$), that every element γ of $\mathcal{G}$ is a side pairing transformation for P, and that there is a side pairing transformation in $\mathcal{G}$ associated to each side of P. Note that by virtue of the definition of a side pairing transformation, if γ is an element of $\mathcal{G}$, then γ^{-1} is also an element of $\mathcal{G}$.

Let $v = v_0$ be a vertex of P, and note that v_0 is an endpoint of two sides of P, which we label as s_0 and s_0'.

Let γ_0 be the side pairing transformation associated to s_0', so that $s_1 = \gamma_0(s_0')$ is another side of P. Note that $v_1 = \gamma_0(v_0)$ is an endpoint of s_1, and of another side of P as well, which we label as s_1'.

Let γ_1 be the side pairing transformation associated to s'_1, so that $s_2 = \gamma_1(s'_1)$ is another side of P. Note that $v_2 = \gamma_1(v_1)$ is an endpoint of s_2, and of another side of P as well, which we label as s'_2.

Continuing this process gives a sequence of side pairing transformations γ_0, $\gamma_1, \ldots$ and a sequence of sides $s_0, s'_0, s_1, s'_1, \ldots$ of P, where s_k and s'_k share a vertex and where $\gamma_k(s'_k) = s_{k+1}$.

Since P has only finitely many sides, there exists some $n \geq 0$ so that $\gamma_n(s'_n) = s_0$. That is, eventually we return to the vertex at which this process began. Set $\gamma = \gamma_n \circ \cdots \circ \gamma_0$. We refer to γ as a *cycle transformation* associated to the vertex v_0. Note that $\gamma(s'_0) = s_0$.

Note that there is another obvious cycle transformation associated to v_0, namely we could have started with the side s_0 at the vertex v_0. The resulting cycle transformation would then be γ^{-1}.

Since the cycle transformation γ at v_0 fixes v_0, either γ is the identity, or γ is elliptic, or γ is reflection in a hyperbolic line. If γ is elliptic, let τ_γ be its order, so that $\tau_\gamma \geq 2$. If γ is the identity, set $\tau_\gamma = 1$, while if γ is reflection, set $\tau_\gamma = 2$.

Let $\alpha(v_k)$ be the interior angle of P at the vertex v_k. We can then associate to v_0 its angle sum

$$\text{sum}(v_0) = \sum_{k=0}^{n} \alpha(v_k).$$

We may view the side pairing transformations as a recipe for gluing together translates of P. The translates P to be glued together around v_0 are

$$\gamma_0(P),\ \gamma_1 \circ \gamma_0(P), \ldots, \gamma_n \circ \cdots \circ \gamma_0(P), \ldots, (\gamma_n \circ \cdots \circ \gamma_0)^{\tau_\gamma}(P).$$

In order for these and their translates under the cycle transformation γ associated to v_0 to glue together to form a neighbourhood of v_0, we need to have that the angle sum at v_0 satisfies

$$\tau_\gamma \,\text{sum}(v_0) = 2\pi.$$

Poincaré's polygon theorem, Theorem 6.19, states that these conditions on the side pairing transformations of P are sufficient to guarantee that the group Γ generated by the side pairing transformations of P is a discrete subgroup of $\text{Möb}(\mathbb{H})$ with P as a fundamental polygon.

Theorem 6.19

Let P be a compact hyperbolic polygon in $\mathbb{H}$, and let $\mathcal{G}$ be a complete collection of side pairing transformations for P. Let v be a vertex of P, and let γ_v be a cycle transformation associated to v of order τ_v. If for each vertex v of P the angle sum $\mathrm{sum}(v)$ of v satisfies $\tau_v \,\mathrm{sum}(v) = 2\pi$, then the subgroup $\Gamma_{\mathcal{G}}$ of $\mathrm{M\ddot{o}b}(\mathbb{H})$ generated by $\mathcal{G}$ is discrete and has P as a fundamental polygon.

It should be noted that this is not the most general statement of the polygon theorem. The polygon theorem holds for non-compact finite sided hyperbolic polygons, and also for infinite sided hyperbolic polygons as well, though in the case of a non-compact hyperbolic polygon there are some technical complications we have chosen to avoid here. Further reading about Theorem 6.19 can be found in the book of Beardon [2], as well as the article of Epstein and Petronio [7].

We close this section with an example illustrating Theorem 6.19. First, consider the compact hyperbolic triangle P with interior angles $\frac{1}{n}\pi$, $\frac{1}{m}\pi$, and $\frac{1}{p}\pi$, where $n \le m \le p$ are positive integers.

Let s_1 be the side of P joining the vertices with interior angles $\frac{1}{m}\pi$ and $\frac{1}{p}\pi$, let s_2 be the side of P joining the vertices with interior angles $\frac{1}{n}\pi$ and $\frac{1}{p}\pi$, and let s_3 be the side of P joining the vertices with interior angles $\frac{1}{n}\pi$ and $\frac{1}{m}\pi$.

Let ℓ_k be the hyperbolic line in $\mathbb{H}$ containing s_k, and let γ_k be reflection in ℓ_k.

Though it is not essential to this example, we note here that by Theorem 5.15, the hyperbolic area of P is

$$\mathrm{area}_{\mathbb{H}}(P) = \pi - \left(\frac{1}{n}\pi + \frac{1}{m}\pi + \frac{1}{p}\pi\right) = \frac{nmp - nm - np - mp}{nmp}\pi.$$

In order for this to be positive, we have that $n \ge 2$, $m \ge 3$, and $p \ge 7$.

Let v_1 be the vertex with interior angle $\frac{1}{n}\pi$. We determine the cycle transformation φ_1 associated to v_1. The two sides of P adjacent to v_1 are s_2 and s_3. Take s_2. The side pairing transformation associated to s_2 is γ_2, which satisfies $\gamma_2(v_1) = v_1$ and $\gamma_2(s_2) = s_2$.

The other side of P adjacent to v_1 is s_3, and the side pairing transformation associated to s_3 is γ_3, which satisfies $\gamma_3(v_1) = v_2$ and $\gamma_3(s_3) = s_3$.

Since the side other than s_3 adjacent to v_1 is s_2, the side we began with, the cycle transformation φ_1 associated to v_1 is $\varphi_1 = \gamma_3 \circ \gamma_2$. Since γ_2 and γ_3 are reflections in hyperbolic lines that intersect with angle $\frac{1}{n}\pi$, their composition φ_1 is an elliptic element of $\mathrm{M\ddot{o}b}(\mathbb{H})$ that is rotation by $\frac{2}{n}\pi$. In particular, the order of φ_1 is $\tau_{\varphi_1} = n$.

We now need to calculate the angle sum $\mathrm{sum}(v_1)$ at v_1. That is, we need to take the sum of the interior angles of P at $\gamma_2(v_1) = v_1$ and $\gamma_3 \circ \gamma_2(v_1) = v_1$.

That is,

$$\mathrm{sum}(v_1) = \frac{1}{n}\pi + \frac{1}{n}\pi = \frac{2}{n}\pi,$$

and so

$$r_{\varphi_1}\,\mathrm{sum}(v_1) = n\,\frac{2}{n}\pi = 2\pi,$$

as desired.

The same argument applies to the other two vertices of P as well. Since the conditions of Theorem 6.19 are satisfied, the subgroup Γ of $\mathrm{M\ddot{o}b}(\mathbb{H})$ generated by γ_1, γ_2, and γ_3 is discrete and has P as a fundamental polygon.

Exercise 6.13

Let P be the regular hyperbolic octagon in $\mathbb{D}$ with interior angle $\frac{1}{4}\pi$, with the side pairings as described earlier in this section. Use Theorem 6.19 to show that the group generated by $\gamma_1, \ldots, \gamma_4$ is a discrete subgroup of $\mathrm{M\ddot{o}b}(\mathbb{H})$.

Solutions

Solutions to Chapter 1 exercises:

1.1: Write $x = \mathrm{Re}(z) = \frac{1}{2}(z + \overline{z})$ and $y = \mathrm{Im}(z) = \frac{-i}{2}(z - \overline{z})$, so that

$$ax + by + c = a\frac{1}{2}(z + \overline{z}) + b\frac{-i}{2}(z - \overline{z}) + c = \frac{1}{2}(a - ib)z + \frac{1}{2}(a + bi)\overline{z} + c = 0.$$

Note that the slope of this line is $-\frac{a}{b}$, which is the quotient of the imaginary and real parts of the coefficient of z.

Given the circle $(x-h)^2 + (y-k)^2 = r^2$, set $z_0 = h + ik$ and rewrite the equation of the circle as

$$|z - z_0|^2 = z\overline{z} - \overline{z_0}z - z_0\overline{z} + |z_0|^2 = r^2.$$

1.2: A and $\mathbb{S}^1$ are perpendicular if and only if their tangent lines at the point of intersection are perpendicular. Let x be a point of $A \cap \mathbb{S}^1$, and consider the Euclidean triangle T with vertices 0, $re^{i\theta}$, and x. The sides of T joining x to the other two vertices are radii of A and $\mathbb{S}^1$, and so A and $\mathbb{S}^1$ are perpendicular if and only if the interior angle of T at x is $\frac{1}{2}\pi$, which occurs if and only if the Pythagorean theorem holds, which occurs if and only if $s^2 + 1^2 = r^2$.

1.3: Let L_{pq} be the Euclidean line segment joining p and q. The midpoint of L_{pq} is $\frac{1}{2}(p+q)$ and the slope of L_{pq} is $m = \frac{\mathrm{Im}(q) - \mathrm{Im}(p)}{\mathrm{Re}(q) - \mathrm{Re}(p)}$. The perpendicular bisector K of L_{pq} passes through $\frac{1}{2}(p+q)$ and has slope $-\frac{1}{m} = \frac{\mathrm{Re}(p) - \mathrm{Re}(q)}{\mathrm{Im}(q) - \mathrm{Im}(p)}$, and so K has the equation

$$\begin{aligned} & y - \frac{1}{2}(\mathrm{Im}(p) + \mathrm{Im}(q)) \\ = & \left[\frac{\mathrm{Re}(p) - \mathrm{Re}(q)}{\mathrm{Im}(q) - \mathrm{Im}(p)}\right]\left(x - \frac{1}{2}(\mathrm{Re}(p) + \mathrm{Re}(q))\right). \end{aligned}$$

The Euclidean centre c of A is the x-intercept of K, which is

$$\begin{aligned} c &= \left[-\frac{1}{2}(\mathrm{Im}(p)+\mathrm{Im}(q))\right]\left[\frac{\mathrm{Im}(q)-\mathrm{Im}(p)}{\mathrm{Re}(p)-\mathrm{Re}(q)}\right] \\ &\quad +\frac{1}{2}(\mathrm{Re}(p)+\mathrm{Re}(q)) \\ &= \frac{1}{2}\left[\frac{(\mathrm{Im}(p))^2-(\mathrm{Im}(q))^2+(\mathrm{Re}(p))^2-(\mathrm{Re}(q))^2}{\mathrm{Re}(p)-\mathrm{Re}(q)}\right] \\ &= \frac{1}{2}\left[\frac{|p|^2-|q|^2}{\mathrm{Re}(p)-\mathrm{Re}(q)}\right]. \end{aligned}$$

The Euclidean radius of A is

$$r = |c-p| = \left|\frac{1}{2}\left[\frac{|p|^2-|q|^2}{\mathrm{Re}(p)-\mathrm{Re}(q)}\right]-p\right|.$$

1.4: One hyperbolic line through i that is parallel to ℓ is the positive imaginary axis $I=\mathbb{H}\cap\{\mathrm{Re}(z)=0\}$. To get a second hyperbolic line through i and parallel to ℓ, take any point x on $\mathbb{R}$ between 0 and 3, say $x=2$, and consider the Euclidean circle centred on $\mathbb{R}$ through 2 and i.

By Exercise 1.3, the Euclidean centre c of A is $c=\frac{3}{4}$ and the Euclidean radius of A is $r=\frac{5}{4}$. Since the real part of every point on A is at most 2, the hyperbolic line $\mathbb{H}\cap C$ is a hyperbolic line passing through i that is parallel to ℓ.

1.5: The Euclidean circle D through i and concentric to A has Euclidean centre -2 and Euclidean radius $\sqrt{5}=|i-(-2)|$, and so one hyperbolic line through i parallel to A is $\mathbb{H}\cap D$.

To construct a second hyperbolic line through i parallel to ℓ, start by taking a point x on $\mathbb{R}$ between A and D, say $x=-4$. Let E be the Euclidean circle centred on $\mathbb{R}$ passing through -4 and i. By Exercise 1.3, the Euclidean centre c of E is $c=-\frac{15}{8}$ and the Euclidean radius is $r=\frac{17}{8}$.

It is an easy calculation that the two Euclidean circles $\{|z+2|=1\}$ and $\{\left|z+\frac{15}{8}\right|=\frac{17}{8}\}$ are disjoint, and so the hyperbolic line $\mathbb{H}\cap E$ is a hyperbolic line passing through i that is parallel to ℓ.

1.6: If $c=0$, then the Euclidean line L_c passing through i and 0 intersects $\mathbb{S}^1$ at $\pm i$, and so $\xi^{-1}(0)=-i$.

Given a point $c\neq 0$ in $\mathbb{R}$, the equation of the Euclidean line L_c passing through c and i is

$$y=-\frac{1}{c}(x-c)=-\frac{1}{c}x+1.$$

To find where L_c intersects $\mathbb{S}^1$, we find the values of x for which

$$|x+i\,y|=\left|x+i\left(-\frac{1}{c}x+1\right)\right|=1,$$

which simplifies to

$$x\left[\left(1+\frac{1}{c^2}\right)x-\frac{2}{c}\right]=0.$$

Since $x=0$ corresponds to i, we have that

$$x=\frac{2c}{c^2+1}.$$

So,

$$\xi^{-1}(c)=\frac{2c}{c^2+1}+i\,\frac{1-c^2}{c^2+1}.$$

1.7: Calculating,

$$\xi(1)=1;\ \xi\left(\exp\left(\frac{2\pi}{3}i\right)\right)=\frac{1}{\sqrt{3}-2};\ \text{ and }\xi\left(\exp\left(\frac{4\pi}{3}i\right)\right)=\frac{-1}{\sqrt{3}+2}.$$

1.8: Let z be a point of $\mathbb{H}$. The Euclidean distance from z to $\mathbb{R}$ is $\text{Im}(z)$. So, $U_{\text{Im}(z)}(z)$ is contained in $\mathbb{H}$, but $U_\varepsilon(z)$ is not contained in $\mathbb{H}$ for any $\varepsilon > \text{Im}(z)$.

1.9: Recall that K is bounded if and only if there exists some $M>0$ so that K is contained in $U_M(0)$. In particular, we have that $U_M(\infty)$ is contained in X. For any $z\in\mathbb{C}-K$, the Euclidean distance $\delta(z)$ from z to K is positive, since K is closed, and so $U_\varepsilon(z)$ is contained in X for any $0<\varepsilon<\delta(z)$. Hence, the complement X of K in $\overline{\mathbb{C}}$ is open.

Suppose that W is an open subset of $\overline{\mathbb{C}}$. If $\infty\notin W$, then W is contained in $\mathbb{C}$, and by the definition of $U_\varepsilon(z)$ we have that W is open in $\mathbb{C}$.

If $\infty\in W$, then $U_\varepsilon(\infty)$ is contained in W for some $\varepsilon>0$. For this same choice of ε, we have that the complement $Y=\overline{\mathbb{C}}-W$ of W is contained in $U_\varepsilon(0)$, and so Y is bounded. The fact that Y is closed follows immediately from the fact that its complement $\overline{\mathbb{C}}-Y=W$ is open in $\overline{\mathbb{C}}$.

1.10: Given $\varepsilon>0$, we need to find $N>0$ so that $z_n=\frac{1}{n}\in U_\varepsilon(0)$ for $n>N$. Take $N=\frac{1}{\varepsilon}$. Then, for $n>N$, we have that $z_n=\frac{1}{n}<\varepsilon$, as desired.

Given $\varepsilon>0$, we need to find $N>0$ so that $w_n=n\in U_\varepsilon(\infty)$ for $n>N$. Take $N=\varepsilon$. Then, for $n>N$, we have that $w_n=n>\varepsilon$, and so $w_n\in U_\varepsilon(\infty)$, as desired.

1.11: Note that 0 lies in $\overline{X}$, since $\frac{1}{n}\in U_\varepsilon(0)\cap X$ for every $n>\varepsilon$. However, there are no other points of $\overline{X}$ other than 0 and the points of X.

If $z\in\mathbb{C}$ is any point with $\text{Im}(z)\neq 0$, then $U_{\text{Im}(z)}(z)\cap X=\emptyset$. Also, since $|x|\leq 1$ for every $x\in X$, we see that $U_2(\infty)\cap X=\emptyset$.

If $z \in \mathbb{R}$ is any point with $\mathrm{Re}(z) \neq 0$ and $\mathrm{Re}(z) \neq \frac{1}{n}$ for $n \in \mathbb{Z} - \{0\}$, then z lies between $\frac{1}{m}$ and $\frac{1}{p}$ for some m, $p \in \mathbb{Z} - \{0\}$. Let $\varepsilon = \min\left(\left|z - \frac{1}{m}\right|, \left|z - \frac{1}{m}\right|\right)$, so that $U_\varepsilon(z) \cap X = \emptyset$.

Hence, $\overline{X} = X \cup \{0\}$.

For Y, take any $z = x + y\,i \in \mathbb{C}$. Given any $\varepsilon > 0$, there exist rational numbers a, b so that $|x - a| < \frac{1}{2}\varepsilon$ and $|y - b| < \frac{1}{2}\varepsilon$, since $\mathbb{Q}$ is dense in $\mathbb{R}$, a fact that can be proven by considering decimal expansions.

Then, $|(x + y\,i) - (a + b\,i)| < \varepsilon$. Since for each $\varepsilon > 0$ we can construct a point in $U_\varepsilon(z) \cap Y$, we have that every point of $\mathbb{C}$ lies in $\overline{Y}$.

Since for any $\varepsilon > 0$ we have that $n \in U_\varepsilon(\infty)$ for every integer n with $n > \varepsilon$, we also have that $\infty \in \overline{Y}$.

Hence, $\overline{Y} = \overline{\mathbb{C}}$.

1.12: In order to show that $\overline{X}$ is closed in $\overline{\mathbb{C}}$, we show that $\overline{\mathbb{C}} - \overline{X}$ is open in $\overline{\mathbb{C}}$. Take $z \in \overline{\mathbb{C}} - \overline{X}$.

Suppose that for each $\varepsilon > 0$, the intersection $U_\varepsilon(z) \cap \overline{X} \neq \emptyset$. For each $n \in \mathbb{N}$, choose some $z_n \in U_{1/n}(z) \cap \overline{X}$. Since $z_n \in \overline{X}$, there is some $x_n \in X$ so that $x_n \in U_{1/n}(z_n) \cap X$.

Combining these yields that $|x_n - z| \leq |x_n - z_n| + |z_n - z| < \frac{2}{n}$. Hence, for each $n \in \mathbb{N}$, we have that $x_n \in U_{2/n}(z) \cap X$, which implies that $z \in \overline{X}$. This contradicts our original choice of z.

1.13: The Euclidean line L_{P} can be expressed parametrically as

$$\mathrm{N} + t(\mathrm{P} - \mathrm{N}) = (tp_1, tp_2, tp_3 + 1 - t)$$

for $t \in \mathbb{R}$. L_{P} intersects the x_1x_2-plane when $tp_3 + 1 - t = 0$, that is when $t = \frac{1}{1 - p_3}$. Hence, we see that

$$\xi(\mathrm{P}) = \frac{p_1}{1 - p_3} + i\frac{p_2}{1 - p_3}.$$

For ξ^{-1}, let $z = x + iy$ be any point of $\mathbb{C}$, and note that z corresponds to the point $Z = (x, y, 0)$ in $\mathbb{R}^3$. Let L be the Euclidean line between N and Z, and note that L is given parametrically by

$$\mathrm{N} + t(Z - \mathrm{N}) = (tx, ty, 1 - t)$$

for $t \in \mathbb{R}$. In order to find where L intersects $\mathbb{S}^2$, we find the point on L whose distance from the origin is 1, which involves solving

$$(tx)^2 + (ty)^2 + (1 - t)^2 = t^2|z|^2 + t^2 - 2t + 1 = 1$$

for t. There are two solutions: $t = 0$, which corresponds to N, and $t = \frac{2}{1+|z|^2}$. The latter value of t yields

$$\xi^{-1}(z) = \left(\frac{2\,\mathrm{Re}(z)}{|z|^2+1}, \frac{2\,\mathrm{Im}(z)}{|z|^2+1}, \frac{|z|^2-1}{|z|^2+1} \right).$$

1.14: Write

$$g(z) = a_n z^n + \cdots + a_1 z + a_0,$$

for $n \geq 1$ with $a_n \neq 0$. We need to quantify the statement that, if $|z|$ is large, then $|g(z)|$ is large. If we wanted to be precise, we could proceed as follows. By the triangle inequality,

$$|g(z)| \geq |\, |a_n z^n| - |a_{n-1} z^{n-1} + \cdots + a_1 z + a_0| \,|.$$

So, set $A = \max\{|a_{n-1}|, \ldots, |a_0|\}$ and note that

$$|a_{n-1} z^{n-1} + \cdots + a_1 z + a_0| \leq A \left(|z^{n-1}| + \cdots + |z| + 1\right) \leq nA|z|^{n-1}$$

for $|z| \geq 1$.

So, given $\varepsilon > 0$, choose $\delta > 0$ so that $\delta > 1$ and so that $|a_n|\delta^n - nA\delta^{n-1} > \varepsilon$. Then, for $|z| > \delta$ we have that

$$\begin{aligned} |g(z)| &\geq |a_n||z|^n - |a_{n-1} z^{n-1} + \cdots + a_1 z + a_0| \\ &\geq |a_n||z|^n - nA|z|^{n-1} \\ &\geq \delta^{n-1}(|a_n|\delta - nA) > \varepsilon, \end{aligned}$$

as desired.

Note that for the constant function $h : \overline{\mathbb{C}} \to \overline{\mathbb{C}}$ defined by $h(z) = c$ for all $z \in \overline{\mathbb{C}}$, we may take $\delta = 1$ for any choice of $\varepsilon > 0$ to see that h is continuous.

1.15: If $d = \mathrm{degree}(g) \geq 2$, then f is not a bijection. In fact, the fundamental theorem of algebra gives that there is a point c of $\overline{\mathbb{C}}$ so that there are at least two distinct solutions to $g(z) = c$. If $g(z)$ does not factor as $g(z) = (z-a)^d$, then we may take $c = 0$.

If $g(z) = (z-a)^d$, then we may take $c = 1$, so that the solutions to $g(z) = 1$ are

$$\left\{ z = a + \exp\left(\frac{2\pi k}{d} i\right) \mid 0 \leq k \leq d \right\}.$$

If $d = \mathrm{degree}(g) = 0$, then f is a constant function, which is continuous but is not a bijection.

If $d = \mathrm{degree}(g) = 1$, then $g(z) = az + b$ where $a \neq 0$. We know from Exercise 1.14 that f is continuous. To see that f is a bijection and that f^{-1} is continuous, we write down an explicit expression for f^{-1}, namely

$$f^{-1}(z) = \frac{1}{a}(z - b) \text{ for } z \in \mathbb{C} \text{ and } f^{-1}(\infty) = \infty.$$

1.16: Choose any $z \in \overline{\mathbb{C}}$. Since X is dense in $\overline{\mathbb{C}}$, there exists a sequence $\{x_n\}$ in X converging to z.

Since f is continuous, we know that $\{f(x_n)\}$ converges to $f(z)$. Since $f(x_n) = x_n$, this gives that $\{x_n\}$ converges to both z and $f(z)$, and so $z = f(z)$.

1.17: Let C_k be the circle in $\overline{\mathbb{C}}$ containing ℓ_k. There are several cases to consider, depending on whether C_1 and C_2 are both Euclidean circles, whether both are Euclidean lines, or whether one is a Euclidean line and one is a Euclidean circle.

We give a complete answer for the case that both C_1 and C_2 are Euclidean circles; the other two cases follow by a similar argument. In essence, we are reproving the fact that if C_1 and C_2 are two Euclidean circles in $\mathbb{C}$, then there exists a circle perpendicular to both C_1 and C_2 if and only if C_1 and C_2 are disjoint.

Let c_k be the Euclidean centre of C_k and let r_k be its Euclidean radius. Suppose there exists a hyperbolic line ℓ that is perpendicular to both ℓ_1 and ℓ_2. Let A be the circle in $\overline{\mathbb{C}}$ containing ℓ.

It may be that A is a Euclidean line, in which case C_1 and C_2 are concentric, and so are ultraparallel, since the boundary at infinity of the closed region in $\mathbb{H}$ bounded by ℓ_1 and ℓ_2 is the union of two closed arcs in $\overline{\mathbb{R}}$.

Otherwise, A is a Euclidean circle with Euclidean centre c and Euclidean radius r. Then, A intersects both C_1 and C_2 perpendicularly, and so $|c - c_k|^2 = r^2 + r_k^2$ by the Pythagorean theorem. In particular, $|c - c_k| > r_k$. Hence, the boundary at infinity of the closed region in $\mathbb{H}$ bounded by ℓ_1 and ℓ_2 consists of two closed arcs, one containing c and the other containing ∞.

Conversely, suppose that ℓ_1 and ℓ_2 are ultraparallel, let H be the closed region in $\mathbb{H}$ bounded by them, and let a be the closed arc in the boundary at infinity of H not containing ∞.

Let C_k be the Euclidean circle containing ℓ_k, let c_k be the Euclidean centre of C_k and let r_k be the Euclidean radius of C_k. For each $x \in a$, consider the Euclidean circle A with Euclidean centre x and Euclidean radius r. In order for A to be perpendicular to C_k, we need that $(c_k - x)^2 = r_k^2 + r^2$.

Solving for x gives

$$x = \frac{r_1^2 - r_2^2 + c_2^2 - c_1^2}{2(c_2 - c_1)}.$$

Since C_1 and C_2 are disjoint Euclidean circles, we have that $c_2 - c_1 > r_1 + r_2$. Hence,

$$r = \sqrt{(x - c_1)^2 - r_1^2} = \sqrt{(x - c_2)^2 - r_2^2} > 0,$$

and so the hyperbolic line contained in A is perpendicular to both ℓ_1 and ℓ_2.

1.18: Suppose that $q = \infty$. Then, the hyperbolic line contained in the Euclidean line $\{\mathrm{Re}(z) = \mathrm{Re}(p)\}$ is the unique hyperbolic line whose endpoints at infinity are p and q. A similiar argument holds if $p = \infty$.

Suppose that $p \neq \infty$ and that $q \neq \infty$. Then, we may again use the construction from the proof of Proposition 1.2 of the perpendicular bisector of the Euclidean line segment joining p to q to find the unique Euclidean circle centred on the real axis $\mathbb{R}$ that passes through both p and q. Intersecting this circle with $\mathbb{H}$ yields the unique hyperbolic line determined by p and q.

Solutions to Chapter 2 exercises:

2.1: Consider the function $f : \overline{\mathbb{C}} \to \overline{\mathbb{C}}$ given by

$$f(z) = \begin{cases} z & \text{for } \mathrm{Re}(z) \leq 0; \\ z + i\,\mathrm{Re}(z) & \text{for } \mathrm{Re}(z) \geq 0; \\ \infty & \text{for } z = \infty. \end{cases}$$

It is evident that f is continuous. To see that f is a bijection and that f^{-1} is continuous, we give an explicit formula for f^{-1}, namely

$$f^{-1}(z) = \begin{cases} z & \text{for } \mathrm{Re}(z) \leq 0; \\ z - i\,\mathrm{Re}(z) & \text{for } \mathrm{Re}(z) \geq 0; \\ \infty & \text{for } z = \infty. \end{cases}$$

Hence, we see that $f \in \mathrm{Homeo}(\overline{\mathbb{C}})$. However, the image of $\overline{\mathbb{R}}$ under f is not a circle in $\overline{\mathbb{C}}$, and so $f \notin \mathrm{Homeo^C}(\overline{\mathbb{C}})$.

2.2: We follow the argument given in the proof of Proposition 2.1. Set $w = az + b$, so that $z = \frac{1}{a}(w - b)$. Substituting this into the equation of a Euclidean circle, namely $\alpha z\overline{z} + \beta z + \overline{\beta}\overline{z} + \gamma = 0$, yields

$$\begin{aligned} \alpha z\overline{z} + \beta z + \overline{\beta}\overline{z} + \gamma &= \alpha \frac{1}{a}(w - b)\overline{\frac{1}{a}(w - b)} + \beta \frac{1}{a}(w - b) + \overline{\beta}\,\overline{\frac{1}{a}(w - b)} + \gamma \\ &= \frac{\alpha}{|a|^2}(w - b)\overline{(w - b)} + \frac{\beta}{a}(w - b) + \overline{\left(\frac{\beta}{a}\right)}\,\overline{w - b} + \gamma \\ &= \frac{\alpha}{|a|^2}\left| w + \frac{\overline{\beta}a}{\alpha} - b \right|^2 + \gamma - \frac{|\beta|^2}{\alpha} = 0, \end{aligned}$$

which is again the equation of a Euclidean circle in $\mathbb{C}$.

2.3: The Euclidean circle A given by the equation $\alpha z\overline{z} + \beta z + \overline{\beta}\overline{z} + \gamma = 0$ has Euclidean centre $-\frac{\overline{\beta}}{\alpha}$ and Euclidean radius $\frac{1}{|\alpha|}\sqrt{|\beta|^2 - \alpha\gamma}$.

As we saw in the solution to Exercise 2.2, the image of A under f is the Euclidean circle given by the equation

$$\frac{\alpha}{|a|^2}\left| w + \frac{\overline{\beta}a}{\alpha} - b \right|^2 + \gamma - \frac{|\beta|^2}{\alpha} = 0,$$

which has Euclidean centre $f\left(-\frac{\overline{\beta}}{\alpha}\right)$ and Euclidean radius $|a|\frac{1}{|\alpha|}\sqrt{|\beta|^2 - \alpha\gamma}$.

2.4: The equation of A expands to

$$z\overline{z} - \overline{z_0}z - z_0\overline{z} + |z_0|^2 - r^2 = 0.$$

The circle $J(C)$ then has the equation

$$(|z_0|^2 - r^2)|w|^2 - z_0 w - \overline{z_0 w} + 1 = 0,$$

which is a Euclidean line if and only if $|z_0|^2 - r^2 = 0$, that is, if and only if $|z_0| = r$.

2.5: The inverse of $m(z) = \frac{az+b}{cz+d}$ is $m^{-1}(z) = \frac{dz-b}{-cz+a}$.

2.6: If all the coefficients of p are zero, then p is undefined. So, write $p(z) = \frac{az+b}{cz+d}$ and suppose that $ad - bc = 0$, so that $ad = bc$. Assume that $a \neq 0$. The proof in the cases that one of the other coefficients is non-zero is similar. Multiply the numerator and denominator of p by a and simplify to get

$$p(z) = \frac{az+b}{cz+d} = \frac{a(az+b)}{a(cz+d)} = \frac{a(az+b)}{acz+bc} = \frac{a(az+b)}{c(az+b)} = \frac{a}{c},$$

and so p is a constant function.

2.7:

1 Since $m(\infty) = \frac{2}{3} \neq \infty$, the fixed points lie in $\mathbb{C}$ and are the roots of $3z^2 - 3z - 5 = 0$, which are $\frac{1}{6}[3 \pm \sqrt{69}]$.

2 One fixed point is $z = \infty$; the other lies in $\mathbb{C}$ and is the solution to $z = 7z+6$, which is $z = -1$.

3 Since $J(\infty) = 0 \neq \infty$, the fixed points lie in $\mathbb{C}$ and are the roots of $z^2 = 1$, which are $z = \pm 1$.

4 Since $m(\infty) = 1 \neq \infty$, the fixed points lie in $\mathbb{C}$ and are the roots of $z^2 = 0$; in particular, m has only one fixed point, at $z = 0$.

2.8: The general form of the Möbius transformation m taking the triple (∞, z_2, z_3) to the triple $(0, 1, \infty)$ is

$$m(z) = \frac{az+b}{cz+d} = \frac{z_2 - z_3}{z - z_3}.$$

2.9: There are 6, as there are 6 permutations of T. $a(z) = z$ takes $(0, 1, \infty)$ to $(0, 1, \infty)$; $b(z) = -(z-1)$ takes $(0, 1, \infty)$ to $(1, 0, \infty)$; $c(z) = \frac{z}{z-1}$ takes $(0, 1, \infty)$ to $(0, \infty, 1)$; $d(z) = \frac{1}{z}$ takes $(0, 1, \infty)$ to $(\infty, 1, 0)$; $e(z) = \frac{-1}{z-1}$ takes $(0, 1, \infty)$ to $(1, \infty, 0)$; $f(z) = \frac{z-1}{z}$ takes $(0, 1, \infty)$ to $(\infty, 0, 1)$.

2.10: There are many such transformations. One is the Möbius transformation m taking the triple $(i, -1, 1)$ of distinct points on $\mathbb{S}^1 = \partial\mathbb{D}$ to the triple $(0, 1, \infty)$ of distinct points on $\overline{\mathbb{R}} = \partial\mathbb{H}$. Explicitly,

$$m(z) = \frac{z-i}{z-1} \frac{-2}{-1-i}.$$

We still need to check that m takes $\mathbb{D}$ to $\mathbb{H}$, which we do by checking for instance that the imaginary part of $m(0)$ is positive:

$$\text{Im}(m(0)) = \text{Im}\left(\frac{2i}{1+i}\right) = \text{Im}(1+i) > 0.$$

2.11: $f(z) = z^2$ is invariant under Möb^+ if and only if $f(m(z)) = f(z)$ for all $m(z) = \frac{az+b}{cz+d}$ in Möb^+ and all $z \in \overline{\mathbb{C}}$. That is, we need to have that

$$f(m(z)) = \left(\frac{az+b}{cz+d}\right)^2 = z^2,$$

and so

$$c^2z^4 + 2cdz^3 + (d^2 - a^2)z^2 - 2abz - b^2 = 0$$

for all z in $\overline{\mathbb{C}}$. In particular, we have that $c = b = 0$ and that $a^2 = d^2$, and so f is not invariant under Möb^+.

Since $ad - bc = ad = 1$, this gives that either $a = d = \pm 1$ or that $a = -d = i$. In the former case, m is the identity Möbius transformation. In the latter case, $m(z) = -z$. Hence, the only subgroup of Möb^+ under which f is invariant is the subgroup $\langle e(z) = z,\ m(z) = -z\rangle$.

2.12: We proceed by direct calculation. Let $m(z) = \frac{az+b}{cz+d}$ where a, b, c, and d lie in $\mathbb{C}$ and $ad - bc = 1$. Then,

$$\begin{aligned}
[m(z_1), m(z_2); m(z_3), m(z_4)] &= \left[\frac{az_1+b}{cz_1+d}, \frac{az_2+b}{cz_2+d}; \frac{az_3+b}{cz_3+d}, \frac{az_4+b}{cz_4+d}\right] \\
&= \left[\frac{\frac{az_1+b}{cz_1+d} - \frac{az_4+b}{cz_4+d}}{\frac{az_1+b}{cz_1+d} - \frac{az_2+b}{cz_2+d}}\right] \left[\frac{\frac{az_3+b}{cz_3+d} - \frac{az_2+b}{cz_2+d}}{\frac{az_3+b}{cz_3+d} - \frac{az_4+b}{cz_4+d}}\right] \\
&= \left[\frac{(az_1+b)(cz_4+d) - (az_4+b)(cz_1+d)}{(az_1+b)(cz_2+d) - (az_2+b)(cz_1+d)}\right] \\
&\quad \left[\frac{(az_3+b)(cz_2+d) - (az_2+b)(cz_3+d)}{(az_3+b)(cz_4+d) - (az_4+b)(cz_2+d)}\right] \\
&= \left[\frac{z_1 - z_4}{z_1 - z_2}\right] \left[\frac{z_3 - z_2}{z_3 - z_4}\right] \\
&= [z_1, z_2; z_3, z_4].
\end{aligned}$$

2.13: Calculating, we see that

$$\begin{aligned}[2+3i,-2i;1-i,4] &= \left[\frac{2+3i-4}{2+3i+2i}\right]\left[\frac{1-i+2i}{1-i-4}\right] \\ &= \left[\frac{-2+3i}{2+5i}\right]\left[\frac{1+i}{-3-i}\right] \\ &= \left[\frac{11+16i}{29}\right]\left[\frac{-4-2i}{10}\right] = \frac{-12-86i}{290},\end{aligned}$$

which is not real, and so $2+3i$, $-2i$, $1-i$, and 4 do not lie on a circle in $\overline{\mathbb{C}}$.

2.14: Calculating, we see that

$$\begin{aligned}[2+3i,-2i;1-i,s] &= \left[\frac{2+3i-s}{2+3i+2i}\right]\left[\frac{1-i+2i}{1-i-s}\right] \\ &= \left[\frac{2-s+3i}{2+5i}\right]\left[\frac{1+i}{1-s-i}\right] \\ &= \left[\frac{19-2s+(5s-4)i}{29}\right]\left[\frac{-s+(2-s)i}{(s-1)^2+1}\right] \\ &= \frac{(7s^2-33s+8)+(-3s^2-19s+38)i}{29((s-1)^2+1)},\end{aligned}$$

which is real if and only if

$$s = \frac{1}{6}\left[-19 \pm \sqrt{817}\right].$$

Hence, there are exactly two real values of s for which $2+3i$, $-2i$, $1-i$, and s lie on a circle in $\overline{\mathbb{C}}$.

2.15: Calculating, we see that

$$[z_1, z_2; z_3, z_4]_2 = \frac{1}{[z_1, z_2; z_3, z_4]}$$

and

$$[z_1, z_2; z_3, z_4]_3 = \frac{1}{1-[z_1, z_2; z_3, z_4]}.$$

2.16: If n fixes a point x of $\overline{\mathbb{C}}$, then $m = p \circ n \circ p^{-1}$ fixes $p(x)$, since

$$m(p(x)) = (p \circ n \circ p^{-1})(p(x)) = p(n(x)) = p(x).$$

Since $n = p^{-1} \circ m \circ p$, we see conversely that if m fixes a point y of $\overline{\mathbb{C}}$, then n fixes $p^{-1}(y)$. In particular, m and n have the same number of fixed points.

2.17: Since $n_2 \circ n_1^{-1}(0) = 0$ and $n_2 \circ n_1^{-1}(\infty) = \infty$, we can write $n_2 \circ n_1^{-1}(z) = p(z) = cz$ for some $c \in \mathbb{C} - \{0, 1\}$. Hence, $n_2 = p \circ n_1$.

Write $n_k \circ m \circ n_k^{-1}(z) = a_k z$, and note that

$$\begin{aligned} a_2 z = n_2 \circ m \circ n_2^{-1}(z) &= p \circ n_1 \circ m \circ n_1^{-1} \circ p^{-1}(z) \\ &= p \circ (n_1 \circ m \circ n_1^{-1}) \left(\frac{1}{c} z\right) \\ &= p \left(\frac{a_1}{c} z\right) = a_1 z, \end{aligned}$$

and so $a_1 = a_2$, as desired.

2.18: Any Möbius transformation taking x to ∞ and y to 0 can be expressed as $s = J \circ q$, where $J(z) = \frac{1}{z}$ and where q is a Möbius transformation taking x to 0 and y to ∞. Calculating, we see that

$$s \circ m \circ s^{-1}(z) = J \circ (q \circ m \circ q^{-1}) \circ J(z) = \frac{1}{a} z.$$

Since by Exercise 2.17 the multiplier of $q \circ m \circ q^{-1}$ is independent of the actual choice of q, the multiplier of $s \circ m \circ s^{-1}$ is independent of the actual choice of s, subject to the condition that s take x to ∞ and y to 0.

2.19:

1 The fixed points of $m(z) = \frac{2z+5}{3z-1}$ are $z = \frac{1}{6}[3 \pm \sqrt{69}]$. Set

$$q(z) = \frac{z - \frac{1}{6}[3 + \sqrt{69}]}{z - \frac{1}{6}[3 - \sqrt{69}]},$$

and calculate that

$$q \circ m \circ q^{-1}(1) = q \circ m(\infty) = q\left(\frac{2}{3}\right) = \frac{\frac{2}{3} - \frac{1}{6}[3 + \sqrt{69}]}{\frac{2}{3} - \frac{1}{6}[3 - \sqrt{69}]}.$$

So, m is loxodromic.

2 The fixed points of $m(z) = 7z + 6$ are $z = \infty$ and $z = -1$, and so m is either elliptic or loxodromic. Set $q(z) = z + 1$, and calculate that

$$q \circ m \circ q^{-1}(1) = q \circ m(0) = q(6) = 7.$$

So, m is loxodromic.

3 $J(z) = \frac{1}{z}$ has fixed points at ± 1, and so is either elliptic or loxodromic. Instead of conjugating J by a Möbius taking its fixed points to 0 and ∞, we note that $J^2(z) = z$, and so J must be elliptic.

4 The fixed point of $m(z) = \frac{z}{z+1}$ is $z = 0$, and so m is parabolic.

2.20: 1. -34; 2. -1; 3. 2; 4. -4; 5. i; 6. -4.

2.21:

1. $m(z) = \frac{\frac{-2i}{\sqrt{34}}z - \frac{4i}{\sqrt{34}}}{\frac{-5i}{\sqrt{34}}z + \frac{7i}{\sqrt{34}}}$; 2. $J(z) = \frac{i}{iz}$; 3. $m(z) = \frac{\frac{-1}{\sqrt{2}}z - \frac{3}{\sqrt{2}}}{\frac{1}{\sqrt{2}}z + \frac{1}{\sqrt{2}}}$

4. $m(z) = \frac{\frac{1}{2}z - \frac{i}{2}}{\frac{-i}{2}z + \frac{3}{2}}$; 5. $m(z) = \frac{\frac{i\sqrt{2}}{1+i}z + \frac{\sqrt{2}}{1+i}}{\frac{\sqrt{2}}{1+i}}$; 6. $m(z) = \frac{\frac{i}{2}z}{\frac{-i}{2}z - 2i}$.

2.22: Write $m(z) = \frac{az+b}{cz+d}$ and $n(z) = \frac{\alpha z+\beta}{\gamma z+\delta}$. Then,

$$n \circ m(z) = \frac{(\alpha a + \beta c)z + \alpha b + \beta d}{(\gamma a + \delta c)z + \gamma b + \delta d},$$

and

$$m \circ n(z) = \frac{(a\alpha + b\gamma)z + a\beta + b\delta}{(c\alpha + d\gamma)z + c\beta + d\delta}.$$

Hence,

$$\tau(n \circ m) = (\alpha a + \beta c + \gamma b + \delta d)^2 = \tau(m \circ n),$$

as desired.

2.23: Using Exercise 2.22, we see that

$$\tau(p \circ m \circ p^{-1}) = \tau(p^{-1} \circ p \circ m) = \tau(m).$$

2.24: Calculating, we have that $f'(\rho) = 2\rho - 2\rho^{-3} = 2\rho(1 - \rho^{-4})$, and so $f'(\rho) = 0$ if and only if $\rho = 1$. Since

$$\lim_{\rho \to 0+} f(\rho) = \infty = \lim_{\rho \to \infty} f(\rho),$$

we see that $\rho = 1$ is a global minimum. Since $f(1) = 2$, we are done.

2.25:

1 $\tau(m) = \frac{-25}{34}$, and so m is loxodromic with multiplier $\frac{1}{68}\left[-93 - \sqrt{4025}\right]$.

2 $\tau(J) = 0$ and so J is elliptic with multiplier -1.

3 $\tau(m) = 0$, and so m is elliptic with multiplier -1.

4 $\tau(m) = 4$, and so m is parabolic.

5 $\tau(m) = 2$, and so m is elliptic with multiplier i.

6 $\tau(m) = -\frac{9}{4}$, and so m is loxodromic with multiplier -4.

2.26: Instead of calculating, we begin by noting that we can write m as $m(z) = \frac{az+b}{cz+d}$ with $ad - bc = 1$ and $a + d = 2$. Choose p so that $a = 1 + px$ and $d = 1 - px$, and note that this determines p uniquely. Since $ad - bc = 1$, we have that $bc = -p^2x^2$.

The fixed points of m satisfy the equation $(1 + px)z + b = z(cz + (1 - px))$, and so $cz^2 - 2pxz - b = 0$. Completing the square, this becomes $\left(z - \frac{px}{c}\right)^2 = 0$.

Since $z = x$ is one solution, we see that $\frac{px}{c} = x$, and so $c = p$. Since $bc = -p^2x^2$, this yields that $b = -px^2$, as desired.

2.27: Let $n(z) = az$, and let $p(z)$ be a Möbius transformation taking x to 0 and taking y to ∞. For instance, we may take $p(z) = \frac{z-x}{z-y}$. The determinant of p is $\beta^2 = x - y$, and so normalizing we get that

$$p(z) = \frac{\frac{1}{\beta}z - \frac{x}{\beta}}{\frac{1}{\beta}z - \frac{y}{\beta}}.$$

Since m fixes x and y and has multiplier a, we have that $p \circ m \circ p^{-1} = n$, and so $m = p^{-1} \circ n \circ p$. Calculating, we see that

$$p^{-1} \circ n \circ p(z) = \frac{1}{\beta^2}\left[\frac{(x-ay)z + xy(a-1)}{(1-a)z + ax - y}\right] = \frac{\left(\frac{x-ya}{x-y}\right)z + \frac{xy(a-1)}{x-y}}{\left(\frac{1-a}{x-y}\right)z + \frac{xa-y}{x-y}},$$

as desired.

2.28: Obviously, every element $k = \lambda \mathrm{I}$ of K is in $\ker(\mu)$, since $\mu(k)$ is the Möbius transformation $m(z) = \frac{\lambda z}{\lambda} = z$.

Suppose that M is an element of $\mathrm{GL}_2(\mathbb{C})$ so that

$$\mu\left(M = \begin{pmatrix} a & b \\ c & d \end{pmatrix}\right) = \left(m(z) = \frac{az+b}{cz+d}\right)$$

is the identity Möbius transformation. Since $m(0) = 0$, we have that $b = 0$; since $m(\infty) = \infty$, we have that $c = 0$. Since $m(1) = \frac{a}{d} = 1$, we have that $a = d$, and so

$$M = \begin{pmatrix} a & b \\ c & d \end{pmatrix} = a\,\mathrm{I},$$

as desired.

The fact that $\mathrm{M\ddot{o}b}^+$ and $\mathrm{PGL}_2(\mathbb{C})$ follows immediately from the first isomorphism theorem from group theory.

2.29: By definition, $C(z) = \overline{z}$ fixes every point of $\overline{\mathbb{R}}$, and in particular fixes 0, 1, and ∞. However, since $C(i) = -i \neq i$, we see that $C(z)$ is not the identity, and so cannot be an element of $\mathrm{M\ddot{o}b}^+$.

2.30: Let A be the circle in $\overline{\mathbb{C}}$ given by the equation $\alpha z\overline{z} + \beta z + \overline{\beta}\overline{z} + \gamma = 0$. Set $w = C(z) = \overline{z}$, so that $z = \overline{w}$, and note that w then satisfies the equation $\alpha w\overline{w} + \overline{\beta}w + \beta\overline{w} + \gamma = 0$, which is again the equation of a circle in $\overline{\mathbb{C}}$, as desired.

2.31: We check that all possible compositions of pairs again have one of the two desired forms. We already have that the composition of two Möbius transformations is again a Möbius transformation.

We begin by noting that the composition $m \circ C$, where $m(z) = \frac{az+b}{cz+d}$, is

$$(m \circ C)(z) = m(\overline{z}) = \frac{a\overline{z}+b}{c\overline{z}+d}.$$

The composition $m \circ n$, where $n(z) = \frac{\alpha\overline{z}+\beta}{\gamma\overline{z}+\delta}$, is

$$(m \circ n)(z) = \frac{(a\alpha + b\gamma)\overline{z} + a\beta + b\delta}{(c\alpha + d\gamma)\overline{z} + c\beta + d\delta},$$

and so has the desired form. Similarly, the composition $n \circ m$ has the desired form.

The composition $p \circ n$, where $p(z) = \frac{a\overline{z}+b}{c\overline{z}+d}$, is

$$(p \circ n)(z) = \frac{(a\overline{\alpha} + b\overline{\gamma})z + a\overline{\beta} + b\overline{\delta}}{(c\overline{\alpha} + d\overline{\gamma})z + c\overline{\beta} + d\overline{\delta}},$$

and so has the desired form.

2.32: One is

$$p(z) = m(z) = \frac{\frac{1}{\sqrt{2}}z + \frac{i}{\sqrt{2}}}{\frac{i}{\sqrt{2}}z + \frac{1}{\sqrt{2}}},$$

for which we have already seen that $p \circ C \circ p^{-1}(z) = \frac{1}{\overline{z}}$.

Consider also the Möbius transformation n taking $(0, 1, \infty)$ to $(i, 1, -1)$, namely

$$n(z) = \frac{\frac{1-i}{2}z + i}{\frac{-1+i}{2}z + 1},$$

and so

$$(n \circ C \circ n^{-1})(z) = n\left(\frac{\overline{z} + i}{\frac{1+i}{2}\overline{z} + \frac{1+i}{2}}\right) = \frac{1}{\overline{z}},$$

as desired.

2.33: Since $f(z) = az + b$ is the composition of $L(z) = az$ and $P(z) = z + b$, it suffices to check that Proposition 2.20 holds for the transformations L and P.

For P, write $b = \beta e^{i\varphi}$, so that the Euclidean line ℓ passing through 0 and b makes angle φ with $\mathbb{R}$. We express translation along ℓ as the reflection in two lines A and B perpendicular to ℓ, with A passing through 0 and B passing through $\frac{1}{2}b$.

Set $\theta = \varphi - \frac{1}{2}\pi$. Reflection in A is given as

$$C_A(z) = e^{2i\theta}\overline{z} = -e^{2i\varphi}\overline{z},$$

and reflection in B is given as

$$C_B(z) = -e^{2i\varphi}\left(\overline{z} - \frac{1}{2}\overline{b}\right) + \frac{1}{2}b.$$

Calculating,

$$(C_B \circ C_A)(z) = C_B(-e^{2i\varphi}\overline{z}) = -e^{2i\varphi}\left(-e^{-2i\varphi}z - \frac{1}{2}\overline{b}\right) + \frac{1}{2}b = z + b.$$

For L, write $a = \alpha^2 e^{2i\theta}$, and note that L is the composition of $D(z) = \alpha^2 z$ and $E(z) = e^{2i\theta}z$.

We can express D as the composition of the reflection $c(z) = \frac{1}{\overline{z}}$ in $\mathbb{S}^1$ and the reflection $c_2(z) = \frac{\alpha}{\overline{z}}$ in the Euclidean circle with Euclidean centre 0 and Euclidean radius α.

We can express E as the composition of the reflection $C(z) = \overline{z}$ in $\mathbb{R}$ and the reflection $C_2(z) = e^{i\theta}\overline{z}$ in the Euclidean line through 0 making angle θ with $\mathbb{R}$.

2.34: We use the notation of the proof of Theorem 2.23. Since X_k passes through z_0 and z_k, we have that $C(X_k)$ passes through $C(z_0) = \overline{z_0}$ and $C(z_k) = \overline{z_k}$, and so $C(X_k)$ has slope

$$S_k = \frac{\operatorname{Im}(\overline{z_k} - \overline{z_0})}{\operatorname{Re}(\overline{z_k} - \overline{z_0})} = -\frac{\operatorname{Im}(z_k - z_0)}{\operatorname{Re}(z_k - z_0)} = -s_k.$$

The angle $\operatorname{angle}(C(X_1), C(X_2))$ between $C(X_1)$ and $C(X_2)$ is then

$$\begin{aligned}\operatorname{angle}(C(X_1), C(X_2)) &= \arctan(S_2) - \arctan(S_1) \\ &= -\arctan(s_2) + \arctan(s_1) = -\operatorname{angle}(X_1, X_2).\end{aligned}$$

Hence, C is conformal, as it preserves the absolute value of the angle between Euclidean lines.

2.35: In the case that $a = 0$, the condition that $ad - bc = 1$ yields that $c \neq 0$. Consider the two points $m(1) = \frac{b}{c+d}$ and $m^{-1}(\infty) = -\frac{d}{c}$.

Solving for d and b in terms of c, we get

$$d = -m^{-1}(\infty)c \text{ and } b = m(1)(c+d) = (m(1) - m^{-1}(\infty))c.$$

Hence,

$$1 = ad - bc = (m^{-1}(\infty) - m(1))c^2,$$

and so again b, c, and d are either all real or all purely imaginary.

In the case that $c = 0$, we have that $a \neq 0$ and $d \neq 0$. In this case, we can write $m(z) = \frac{a}{d}z + \frac{b}{d}$, and so both $m(0) = \frac{b}{d}$ and $m(1) = \frac{a+b}{d}$ are real. This gives

$$b = m(0)d \text{ and } a = (m(1) - m(0))d.$$

Hence,

$$1 = ad - bc = (m(1) - m(0))d^2,$$

and so again a, b, and d are either all real or all purely imaginary.

2.36: Start by taking an element p of Möb taking $\overline{\mathbb{R}}$ to $\mathbb{S}^1$, such as $p(z) = \frac{z-i}{-iz+1}$. Set $m(z) = \frac{az+b}{cz+d}$ and calculate that

$$p \circ m \circ p^{-1}(z) = \frac{(a+d+(b-c)i)z + b + c + (a-d)i}{(b+c-(a-d)i)z + a + d - (b-c)i}.$$

Set $\alpha = a + d + (b-c)i$ and $\beta = b + c + (a-d)i$.

If a, b, c, and d are all real, then with α and β as above we can rewrite $p \circ m \circ p^{-1}$ as

$$p \circ m \circ p^{-1}(z) = \frac{\alpha z + \beta}{\overline{\beta} z + \overline{\alpha}}.$$

If a, b, c, and d are all purely imaginary, then with α and β as above we can rewrite $p \circ m \circ p^{-1}$ as

$$p \circ m \circ p^{-1}(z) = \frac{\alpha z + \beta}{-\overline{\beta} z - \overline{\alpha}}.$$

If $n(z) = \frac{a\overline{z}+b}{c\overline{z}+d}$, then

$$p \circ n \circ p^{-1}(z) = \frac{(a-d-(b+c)i)\overline{z} + b - c - (a+d)i}{(-b+c-(a+d)i)\overline{z} - a + d - (b+c)i}.$$

Set $\delta = a - d - (b+c)i$ and $\gamma = b - c - (a+d)i$.

If a, b, c, and d are all real, then with δ and γ as above we can rewrite $p \circ n \circ p^{-1}$ as

$$p \circ n \circ p^{-1}(z) = \frac{\delta \overline{z} + \gamma}{-\overline{\gamma}\overline{z} - \overline{\delta}}.$$

If a, b, c, and d are all purely imaginary, then with δ and γ as above we can rewrite $p \circ n \circ p^{-1}$ as

$$p \circ n \circ p^{-1}(z) = \frac{\delta \overline{z} + \gamma}{\overline{\gamma}\overline{z} + \overline{\delta}}.$$

2.37: This is very similar to the proof of Theorem 2.4. First, note that the elements listed as generators are all elements of Möb($\mathbb{H}$). Consider the element $m(z) = \frac{az+b}{cz+d}$ of Möb($\mathbb{H}$), where $a, b, c, d \in \mathbb{R}$ and $ad - bc = 1$.

If $c = 0$, then $m(z) = \frac{a}{d}z + \frac{b}{d}$. Since $1 = ad - bc = ad$, we have that $\frac{a}{d} = a^2 > 0$.

If $c \neq 0$, then $m(z) = f(K(g(z)))$, where $g(z) = c^2 z + cd$ and $f(z) = z + \frac{a}{c}$.

For $n(z) = \frac{a\overline{z}+b}{c\overline{z}+d}$, where a, b, c, and d are purely imaginary and $ad - bc = 1$, note that $B \circ n = m$ where $m(z) = \frac{\alpha z+\beta}{\gamma z+\delta}$ is an element of Möb($\mathbb{H}$). Hence, we can write $n = B^{-1} \circ m = B \circ m$.

2.38: We know from Theorem 2.26 that every element of Möb($\mathbb{H}$) either has the form $m(z) = \frac{az+b}{cz+d}$ where a, b, c, $d \in \mathbb{R}$ and $ad - bc = 1$, or has the form $n(z) = \frac{a\overline{z}+b}{c\overline{z}+d}$, where a, b, c, and d are purely imaginary and $ad - bc = 1$.

The Möbius transformation $p(z) = \frac{z-i}{-iz+1}$ takes $\overline{\mathbb{R}}$ to $\mathbb{S}^1$, and takes $\mathbb{H}$ to $\mathbb{D}$ since $p(i) = 0$.

For m, we calculate

$$p \circ m \circ p^{-1}(z) = \frac{(a+d+(b-c)i)z + b + c + (a-d)i}{(b+c-(a-d)i)z + a + d - (b-c)i} = \frac{\alpha z + \beta}{\overline{\beta} z + \overline{\alpha}},$$

where $\alpha = a + d + (b - c)i$ and $\beta = b + c + (a - d)i$.

For n, we calculate

$$p \circ n \circ p^{-1}(z) = \frac{(a-d-(b+c)i)\overline{z} + b - c - (a+d)i}{(-b+c-(a+d)i)\overline{z} - a + d - (b+c)i} = \frac{\delta\overline{z} + \gamma}{\overline{\gamma}\overline{z} + \overline{\delta}},$$

with $\delta = a - d - (b + c)i$ and $\gamma = b - c - (a + d)i$.

2.39: Let ℓ be a hyperbolic line in $\mathbb{H}$. Using Lemma 2.8, it suffices to construct an element of Möb($\mathbb{H}$) that takes ℓ to the positive imaginary axis I in $\mathbb{H}$. One approach is to construct an element of Möb($\mathbb{H}$) taking the endpoints at infinity of ℓ to 0 and ∞, as is done in the solution to Exercise 2.40. We take another approach here.

Choose a point z on ℓ. By Proposition 2.28, there exists an element m of Möb($\mathbb{H}$) with $m(z) = i$. Let φ be the angle between the two hyperbolic lines I and $m(\ell)$, measured from I to $m(\ell)$.

For each θ, the Möbius transformation

$$n_\theta(z) = \frac{\cos(\theta)z - \sin(\theta)}{\sin(\theta)z + \cos(\theta)}$$

lies in Möb($\mathbb{H}$) and fixes i. Also, the angle between I and $n_\theta(I)$ at i, measured from I to $n_\theta(I)$, is θ.

So, if we take $\theta = -\varphi$, we have that $n_\theta(I)$ and $m(\ell)$ are both hyperbolic lines through i that make angle θ with I. Hence, $m(\ell) = n_\theta(I)$, and so $I = n_\theta^{-1} \circ m(\ell)$.

2.40: For any two points $y < x$ in $\mathbb{R}$, the Möbius transformation $m(z) = \frac{z-x}{z-y}$ satisfies $m(x) = 0$ and $m(y) = \infty$. Also, the determinant of m is $x - y > 0$, and so m lies in Möb$^+$($\mathbb{H}$).

For $y = -2$ and $x = 1$, we get $m(z) = \frac{z-1}{z+2}$ as an element of Möb($\mathbb{H}$) taking ℓ to I.

2.41: Let H be the open half-plane $H = \{z \in \mathbb{H} \mid \text{Re}(z) > 0\}$ determined by the positive imaginary axis I. Yet again using Lemma 2.8, given any open half-plane L in $\mathbb{H}$, it suffices to construct an element of Möb($\mathbb{H}$) taking L to H.

Let ℓ be the bounding line for L. By Exercise 2.39, there is an element m of $\text{Möb}(\mathbb{H})$ satisfying $m(\ell) = I$. In particular, m takes the two open half-planes determined by ℓ to the two open half-planes determined by I.

If $m(L) = H$, we are done. If $m(L) \neq H$, then $B \circ m(L) = H$, where $B(z) = -\overline{z}$ is reflection in I, and we are done.

Solutions to Chapter 3 exercises:

3.1: Since the Euclidean distance from z to $\mathbb{S}^1$ is $1-|z|$, we see that $\delta(z) = \frac{1}{1-|z|}$. Parametrize C_r by the path $f : [0, 2\pi] \to \mathbb{D}$ given by $f(t) = re^{it}$, so that $|f(t)| = r$ and $|f'(t)| = |ire^{it}| = r$. Calculating, we see that

$$\begin{aligned} \text{length}(C_r) = \text{length}(f) &= \int_f \frac{1}{1-|z|}|\mathrm{d}z| \\ &= \int_0^{2\pi} \frac{1}{1-|f(t)|}|f'(t)|\mathrm{d}t \\ &= \int_0^{2\pi} \frac{r}{1-r}\mathrm{d}t = \frac{2\pi r}{1-r}. \end{aligned}$$

3.2: On $[0,1]$, we have that $|f(t)| = |t+ti| = \sqrt{2}\,t$ and $|f'(t)| = \sqrt{2}$, while on $[-1,0]$ we have that $|f(t)| = |t-ti| = \sqrt{2t^2} = -\sqrt{2}\,t$ and $|f'(t)| = \sqrt{2}$. So,

$$\begin{aligned} \text{length}(f) = \int_f \frac{1}{1+|z|^2}|\mathrm{d}z| &= \int_{-1}^{1} \frac{1}{1+2t^2}\sqrt{2}\mathrm{d}t \\ &= 2\arctan(\sqrt{2}). \end{aligned}$$

3.3: Parametrize A_λ by the path $f : [-1,1] \to \mathbb{H}$ given by $f(t) = t + i\lambda$. Since $\text{Im}(f(t)) = \lambda$ and $|f'(t)| = 1$, we see that

$$\text{length}(f) = \int_{-1}^{1} \frac{c}{\lambda}\mathrm{d}t = \frac{2c}{\lambda}.$$

B_λ lies on the Euclidean circle with Euclidean centre 0 and Euclidean radius $\sqrt{1+\lambda^2}$. The Euclidean line segment between 0 and $1+i\lambda$ makes angle θ with the positive real axis, where $\cos(\theta) = \frac{1}{\sqrt{1+\lambda^2}}$. So, we can parametrize B_λ by the path $g : [\theta, \pi-\theta] \to \mathbb{H}$ given by $g(t) = \sqrt{1+\lambda^2}\, e^{i\theta}$.

Since $\text{Im}(g(t)) = \sqrt{1+\lambda^2}\,\sin(\theta)$ and $|g'(t)| = \sqrt{1+\lambda^2}$, we see that

$$\text{length}(g) = \int_\theta^{\pi-\theta} c\csc(t)\,\mathrm{d}t = c\,\ln\left[\frac{\sqrt{1+\lambda^2}+1}{\sqrt{1+\lambda^2}-1}\right].$$

3.4: Since $K'(z) = \frac{1}{z^2}$, the condition imposed on $\rho(z)$ is that

$$0 = \mu_K(z) = \rho(z) - \rho(K(z))|K'(z)| = \rho(z) - \rho\left(-\frac{1}{z}\right)\frac{1}{|z|^2}.$$

Substituting $\rho(z) = \frac{c}{\mathrm{Im}(z)}$ and using that

$$\rho\left(-\frac{1}{z}\right) = \rho\left(\frac{-\overline{z}}{|z|^2}\right) = \frac{c|z|^2}{\mathrm{Im}(-\overline{z})} = \frac{c|z|^2}{\mathrm{Im}(z)},$$

we obtain

$$\rho(z) - \rho\left(-\frac{1}{z}\right)\frac{1}{|z|^2} = \frac{c}{\mathrm{Im}(z)} - \frac{c|z|^2}{\mathrm{Im}(z)}\frac{1}{|z|^2} = \frac{c}{\mathrm{Im}(z)} - \frac{c}{\mathrm{Im}(z)} = 0,$$

as desired.

The calculation for $B(z)$ is similar, but can also be performed directly using the definition of length. For a piecewise differentiable path $f : [a,b] \to \mathbb{H}$ given by $f(t) = x(t) + iy(t)$, we have that $B \circ f(t) = -x(t) + iy(t)$.

Hence, we have that $|(B \circ f)'(t)| = |f'(t)|$ and $\mathrm{Im}(B \circ f)(t) = y(t) = \mathrm{Im}(f(t))$, and so

$$\begin{aligned} \mathrm{length}(B \circ f) &= \int_a^b \frac{c}{\mathrm{Im}((B \circ f)(t))}|(B \circ f)'(t)|\mathrm{d}t \\ &= \int_a^b \frac{c}{\mathrm{Im}(f(t))}|f'(t)|\mathrm{d}t = \mathrm{length}(f), \end{aligned}$$

as desired.

3.5: Since $|f_n'(t)| = |1 + int^{n-1}| = \sqrt{1 + n^2 t^{2n-2}}$ and since $\mathrm{Im}(f_n(t)) = t^n + 1$, we have that

$$\mathrm{length}_{\mathbb{H}}(f_n) = \int_{f_n} \frac{1}{\mathrm{Im}(z)}|\mathrm{d}z| = \int_0^1 \frac{\sqrt{1 + n^2 t^{2n-2}}}{1 + t^n}\mathrm{d}t.$$

For $n = 1$, this gives that

$$\mathrm{length}_{\mathbb{H}}(f_1) = \int_0^1 \frac{\sqrt{2}}{1 + t}\mathrm{d}t = \sqrt{2}\,\ln(2).$$

For $n \geq 2$, this integral is more difficult to evaluate explicitly.

3.6: As $n \to \infty$, the curves $\gamma_n = f_n([0,1])$ seems to converge to the curve γ that is the union of the horizontal Euclidean line segment ℓ_1 joining i and $1+i$ and the vertical Euclidean line segment ℓ_2 joining $1+i$ and $1+2i$.

Consequently, we might expect that $\mathrm{length}_{\mathbb{H}}(\gamma_n) \to \mathrm{length}_{\mathbb{H}}(\gamma)$ as $n \to \infty$.

Parametrizing ℓ_1 by $f : [0,1] \to \mathbb{H}$ given by $f(t) = t + i$, we see that

$$\mathrm{length}_{\mathbb{H}}(\ell_1) = \mathrm{length}_{\mathbb{H}}(f) = 1.$$

Parametrizing ℓ_2 by $g : [1,2] \to \mathbb{H}$ given by $g(t) = 1 + ti$, we see that

$$\mathrm{length}_{\mathbb{H}}(\ell_2) = \mathrm{length}_{\mathbb{H}}(g) = \ln(2).$$

Hence, we have that $\text{length}_{\mathbb{H}}(\gamma) = 1 + \ln(2)$.

3.7: Since $g(-1) = g(1) = 2i$, and since g achieves its minimum at 0, the image of $[-1, 1]$ under g is the hyperbolic line segment joining i to $2i$.

The hyperbolic length of g is

$$\text{length}_{\mathbb{H}}(g) = \int_g \frac{1}{\text{Im}(z)}|\text{d}z| = \int_{-1}^{1} \frac{|2t|}{t^2+1}\text{d}t = 2\ln(2).$$

3.8: For each $n \geq 2$, define the numbers

$$2 = \lambda_0 < \lambda_1 < \ldots < \lambda_n = 10$$

by setting

$$\text{d}_{\mathbb{H}}(\lambda_k i, \lambda_{k+1} i) = \frac{1}{n}\text{d}_{\mathbb{H}}(2i, 10i)$$

for $0 \leq k \leq n-1$. Since

$$\text{d}_{\mathbb{H}}(\lambda_k i, \lambda_{k+1} i) = \ln\left[\frac{\lambda_{k+1}}{\lambda_k}\right],$$

we see that

$$\text{d}_{\mathbb{H}}(\lambda_0 i, \lambda_k i) = \ln\left[\frac{\lambda_k}{\lambda_0}\right] = \frac{k}{n}\text{d}_{\mathbb{H}}(2i, 10i).$$

Hence,

$$\ln(\lambda_k) = \frac{k}{n}\ln(5) + \ln(2),$$

and so

$$\lambda_k = 2 \cdot 5^{\frac{k}{n}} i.$$

For example, for $n = 2$, we get that the midpoint of the hyperbolic line segment between $2i$ and $10i$ is $2\sqrt{5}i$.

3.9: By Exercise 1.3, the Eulidean centre of the Euclidean circle containing the hyperbolic line ℓ passing through z_1 and z_2 is

$$c = \frac{1}{2}\left[\frac{|z_1|^2 - |z_2|^2}{\text{Re}(z_1) - \text{Re}(z_2)}\right] = \frac{1}{2}\left[\frac{|z_1|^2 - |z_2|^2}{x_1 - x_2}\right].$$

Setting the Euclidean radius of the Euclidean circle to be $r = |z_1 - c|$, the endpoints at infinity of ℓ are $c - r$ and $c + r$.

Set $m(z) = \frac{z-(c+r)}{z-(c-r)}$. Since the determinant of m is $c + r - (c - r) = 2r > 0$, we have that m lies in $\text{Möb}^+(\mathbb{H})$. Calculating, we see that

$$m(z_1) = \frac{z_1 - (c+r)}{z_1 - (c-r)} \text{ and } m(z_2) = \frac{z_2 - (c+r)}{z_2 - (c-r)},$$

both of which lie on the positive imaginary axis by construction.

Hence,

$$\begin{aligned} \mathrm{d}_{\mathbb{H}}(z_1, z_2) = \mathrm{d}_{\mathbb{H}}(m(z_1), m(z_2)) &= \left| \ln \left[\frac{m(z_2)}{m(z_1)} \right] \right| \\ &= \left| \ln \left[\frac{(z_2 - (c+r))}{(z_2 - (c-r))} \frac{(z_1 - (c-r))}{(z_1 - (c+r))} \right] \right|. \end{aligned}$$

3.10: We use the notation and formula from Exercise 3.9. For $A = i$ and $B = 1 + 2i$, we have $c = 2$ and $r = \sqrt{5}$, and so

$$\mathrm{d}_{\mathbb{H}}(A, B) = \ln \left[\frac{\sqrt{5}+1}{\sqrt{5}-1} \right].$$

For $A = i$ and $C = -1 + 2i$, we have $c = -2$ and $r = \sqrt{5}$, and so

$$\mathrm{d}_{\mathbb{H}}(A, C) = \ln \left[\frac{\sqrt{5}+1}{\sqrt{5}-1} \right].$$

Note that we expect this, since $A = K(A)$ and $C = K(B)$, where $K(z) = -\overline{z}$ is an element of $\mathrm{M\ddot{o}b}(\mathbb{H})$ and hence preserves hyperbolic distance.

For $A = i$ and $D = 7i$, we have that $\mathrm{d}_{\mathbb{H}}(A, D) = \ln(7)$.

For $B = 1 + 2i$ and $C = -1 + 2i$, we have $c = 0$ and $r = \sqrt{5}$, and so

$$\mathrm{d}_{\mathbb{H}}(B, C) = \ln \left[\frac{\sqrt{5}+1}{\sqrt{5}-1} \right].$$

For $B = 1 + 2i$ and $D = 7i$, we have $c = -22$ and $r = \sqrt{533}$, and so

$$\mathrm{d}_{\mathbb{H}}(B, D) = \ln \left[\frac{41+\sqrt{533}}{41-\sqrt{533}} \right].$$

Since $C = K(B)$ and $D = K(D)$ for $K(z) = -\overline{z}$, we have that

$$\begin{aligned} \mathrm{d}_{\mathbb{H}}(C, D) &= \mathrm{d}_{\mathbb{H}}(K(B), K(D)) \\ &= \mathrm{d}_{\mathbb{H}}(B, D) \\ &= \ln \left[\frac{41+\sqrt{533}}{41-\sqrt{533}} \right]. \end{aligned}$$

3.11: If there exists an element q of $\mathrm{M\ddot{o}b}(\mathbb{H})$ taking (z_1, z_2) to (w_1, w_2), then

$$\mathrm{d}_{\mathbb{H}}(w_1, w_2) = \mathrm{d}_{\mathbb{H}}(q(z_1), q(z_2)) = \mathrm{d}_{\mathbb{H}}(z_1, z_2).$$

If on the other hand we have that $\mathrm{d}_{\mathbb{H}}(w_1, w_2) = \mathrm{d}_{\mathbb{H}}(z_1, z_2)$, we proceed as follows. There exists an element m of $\mathrm{M\ddot{o}b}(\mathbb{H})$ taking z_1 to i and taking z_2 to

$e^{d_{\mathbb{H}}(z_1,z_2)}i$; there also exists an element n of Möb($\mathbb{H}$) taking w_1 to i and taking w_2 to $e^{d_{\mathbb{H}}(w_1,w_2)}i$. Note that $m(z_1) = n(w_1) = i$. Since $d_{\mathbb{H}}(w_1, w_2) = d_{\mathbb{H}}(z_1, z_2)$, we have that $m(z_2) = n(w_2)$, and so $q = n^{-1} \circ m$ takes (z_1, z_2) to (w_1, w_2).

3.12: If $f(x) = f(y)$, then $d(f(x), f(y)) = 0$. Hence, $d(x, y) = 0$, and so $x = y$ by the first of the three conditions describing a metric. Hence, f is injective.

To show that f is continuous at x, take some $\varepsilon > 0$. We need to find $\delta > 0$ so that $f(U_\delta(x)) \subset U_\varepsilon(f(z))$. However, since $d(x, y) = d(f(x), f(y))$, we see that if $y \in U_\delta(x)$, then $d(x, y) < \delta$, and so $d(f(x), f(y)) < \delta$, and so $f(y) \in U_\delta(f(x))$. Hence, take $\delta = \varepsilon$.

3.13: We know from our work in Section 2.1 that $f(z) = az$ is a homeomorphism of $\mathbb{C}$ for every $a \in \mathbb{C} - \{0\}$. Since

$$|f(z) - f(w)| = |az - aw| = |a|\,|z - w|,$$

we see that f is an isometry if and only if $|a| = 1$.

3.14: This follows from Proposition 3.13. Suppose that y lies in the hyperbolic line segment ℓ_{xz} joining x to z. Then,

$$d_{\mathbb{H}}(x, y) + d_{\mathbb{H}}(y, z) = d_{\mathbb{H}}(x, z).$$

Since f is a hyperbolic isometry, it preserves hyperbolic distance, and so

$$d_{\mathbb{H}}(f(x), f(y)) + d_{\mathbb{H}}(f(y), f(z)) = d_{\mathbb{H}}(f(x), f(z)).$$

In particular, $f(y)$ lies in the hyperbolic line segment $\ell_{f(x)\,f(z)}$ joining $f(x)$ to $f(z)$, and so

$$f(\ell_{xz}) = \ell_{f(x)\,f(z)}.$$

Since a hyperbolic line can be expressed as a nested union of hyperbolic line segments, we have that hyperbolic isometries take hyperbolic lines to hyperbolic lines.

3.15: Without loss of generality, we suppose $\lambda < \mu$. Write $y = \xi i$ and consider $d_{\mathbb{H}}(x, y) = |\ln(\lambda) - \ln(\xi)|$. As a function of ξ, $g(\xi) = \ln(\lambda) - \ln(\xi)$ is strictly decreasing, since $g'(\xi) = -\frac{1}{\xi}$. Hence, for any number c, there is at most one solution to $g(\xi) = c$.

Hence, for any $c > 0$, there are two solutions y to $d_{\mathbb{H}}(x, y) = c$. One is $y = e^{\ln(\lambda)-c}i$ and the other is $y = e^{\ln(\lambda)+c}i$. Geometrically, one is above $x = \lambda i$ on I and the other is below.

Similarly, there are two solutions y to $d_{\mathbb{H}}(y, z) = c'$, one above z on I and one below. Hence, there can be only one solution to the two equations $d_{\mathbb{H}}(x, y) = c$ and $d_{\mathbb{H}}(y, z) = c'$.

3.16: The point $2w$ lies on the hyperbolic line contained in the Euclidean circle with Euclidean centre 0 and Euclidean radius $2|w|$. Note that the point of intersection of this Euclidean circle with I is $Z(2w) = 2Z(w)$. Since $\mathrm{d}_{\mathbb{H}}(Z(w), w) = \mathrm{d}_{\mathbb{H}}(Z(2w), 2w)$, we can express $2w$ in these new coordinates as

$$(\log(2w), \mathrm{sign}(2w)\, \mathrm{d}_{\mathbb{H}}(Z(2w), 2w)),$$

or equivalently as

$$(\log(w) + \log(2), \mathrm{sign}(w)\, \mathrm{d}_{\mathbb{H}}(Z(w), w)).$$

Hence, we can express the action of $m(z) = 2z$ in these new coordinates as $m(a, b) = (a + \log(2), b)$. That is, we have converted a dilation to a translation.

3.17: If X and Y do not have disjoint closures, then there exists a point x in $\overline{X} \cap \overline{Y}$. Since $x \in \overline{X}$, there exists a sequence $\{x_n\}$ of points of X converging to X, and since $x \in \overline{Y}$, there exists a sequence $\{y_n\}$ of points of Y converging to Y. Since $\mathrm{d}_{\mathbb{H}}$ is continuous, we have that

$$\lim_{n\to\infty} \mathrm{d}_{\mathbb{H}}(x_n, y_n) = \mathrm{d}_{\mathbb{H}}(x, x) = 0.$$

Since $\mathrm{d}_{\mathbb{H}}(X, Y) \le \mathrm{d}_{\mathbb{H}}(x_n, y_n)$ for all n, we have shown that $\mathrm{d}_{\mathbb{H}}(X, Y) = 0$.

Suppose now that $\mathrm{d}_{\mathbb{H}}(X, Y) = 0$. Since

$$\mathrm{d}_{\mathbb{H}}(X, Y) = \inf\{\mathrm{d}_{\mathbb{H}}(x, y) \,:\, x \in X,\ y \in Y\},$$

there exists a sequence of points $\{x_n\}$ of X and a sequence of points $\{y_n\}$ of Y so that $\lim_{n\to\infty} \mathrm{d}_{\mathbb{H}}(x_n, y_n) = 0$.

Since X is compact, there exists a subsequence $\{x_{n_k}\}$ of $\{x_n\}$ that converges to a point x of X. Since $\lim_{n\to\infty} \mathrm{d}_{\mathbb{H}}(x_n, y_n) = 0$, we have that $\lim_{k\to\infty} \mathrm{d}_{\mathbb{H}}(x_{n_k}, y_{n_k}) = 0$, and so $\{y_{n_k}\}$ converges to x as well. Hence, x is a point of $\overline{X} \cap \overline{Y}$, and so X and Y do not have disjoint closures.

3.18: Begin by applying an element of Möb($\mathbb{H}$) so that ℓ lies in the Euclidean circle with Euclidean centre 0 and Euclidean radius 1 and so that $p = \lambda i$ for some $\lambda > 1$. In this case, the unique hyperbolic line through p that is perpendicular to ℓ is the positive imaginary axis I, which intersects ℓ at i.

Using the formula for $\mathrm{d}_{\mathbb{H}}(z_1, z_2)$ given in Section 3.5, and a lot of algebraic massage, we derive that the hyperbolic distance

$$\mathrm{d}_{\mathbb{H}}(e^{i\theta}, \lambda i) = \ln\left[\cosh(\ln(\lambda))\csc(\theta) + \sqrt{\cosh^2(\ln(\lambda))\csc^2(\theta) - 1}\right].$$

The derivative of this function is negative for $0 < \theta \le \frac{\pi}{2}$, and so $\mathrm{d}_{\mathbb{H}}(e^{i\theta}, \lambda i)$ achieves its unique minumum at $\theta = \frac{\pi}{2}$.

In particular, note that this shows the following. Let ℓ be a hyperbolic line, let p be a point in $\mathbb{H}$ not on ℓ, and let a be the point on ℓ satisfying $\mathrm{d}_{\mathbb{H}}(p,a) = \mathrm{d}_{\mathbb{H}}(p,\ell)$. Then, for a point z in ℓ, the hyperbolic distance $\mathrm{d}_{\mathbb{H}}(p,z)$ is monotonically increasing as a function of $\mathrm{d}_{\mathbb{H}}(a,z)$.

3.19: The distance from the point $\rho e^{i\varphi}$ to the positive imaginary axis I is equal to the hyperbolic length of the hyperbolic line segment from $\rho e^{i\varphi}$ to I that meets I perpendicularly, which is the hyperbolic line segment joining $\rho e^{i\varphi}$ and ρi.

To calculate $\mathrm{d}_{\mathbb{H}}(\rho e^{i\varphi}, \rho i)$, we may use for instance Exercise 3.9. The hyperbolic line passing through $\rho e^{i\varphi}$ and ρi lies on the Euclidean circle with Euclidean centre 0 and Euclidean radius ρ, and so

$$\mathrm{d}_{\mathbb{H}}(\rho e^{i\varphi}, \rho i) = \left| \ln\left[\frac{\sin(\varphi)}{1+\cos(\varphi)}\right]\right|.$$

On $(0, \frac{\pi}{2}]$, we have $1+\cos(\varphi) \geq 1$ and $\sin(\varphi) \leq 1$, and so

$$\mathrm{d}_{\mathbb{H}}(\rho e^{i\varphi}, \rho i) = \ln\left[\frac{1+\cos(\varphi)}{\sin(\varphi)}\right].$$

Hence, W_ε is the set of points of $\mathbb{H}$ for which $\mathrm{d}_{\mathbb{H}}(\rho e^{i\varphi}, \rho i) = \varepsilon$ is constant, and so φ is constant. This is a Euclidean ray from the origin, where we take $\varphi = \frac{\pi}{2} - \theta$.

Since $\mathrm{d}_{\mathbb{H}}(\rho e^{i\varphi}, \rho i) = \mathrm{d}_{\mathbb{H}}(\rho e^{i(\pi-\varphi)}, \rho i)$, we see that W_ε has a second component, namely the Euclidean ray from the origin making angle $\frac{\pi}{2} + \theta = \pi - \varphi$ with the positive real axis.

3.20: Use the triple transitivity of the action of Möb($\mathbb{H}$) on $\overline{\mathbb{R}}$ to assume that that the endpoints at infinity of ℓ_0 are 0 and ∞, and that the endpoints at infinity of ℓ_1 are 1 and ∞.

For each $r > 1$, let c_r be the hyperbolic line contained in the Euclidean circle with Euclidean centre 0 and Euclidean radius r, and note that c_r is the unique hyperbolic line through ri that is perpendicular to ℓ_0. Hence, if there is a hyperbolic line perpendicular to both ℓ_0 and ℓ_1, then it will be one of the c_r.

Write the point of intersection of c_r and ℓ_1 as $re^{i\theta}$, and note that $\cos(\theta) = \frac{\sqrt{r^2-1}}{r}$ is non-zero for all $r > 1$. However, θ is also the angle between ℓ_1 and c_r, measured from ℓ_1 to c_r, and so no c_r intersects ℓ_1 perpendicularly.

3.21: By the ordering of the points around $\overline{\mathbb{R}}$, there exists an element of Möb($\mathbb{H}$) taking z_0 to 0, taking z_1 to ∞, taking w_0 to 1, and taking w_1 to $x > 1$.

From our work in Section 2.3, we know that

$$[z_0, w_0; w_1, z_1] = [0, 1; x, \infty] = \frac{x-1}{0-1} = 1 - x,$$

and so

$$1 - [z_0, w_0; w_1, z_1] = x.$$

Calculating, we see that

$$\tanh^2\left[\frac{1}{2}\mathrm{d}_{\mathbb{H}}(\ell_0, \ell_1)\right] = \tanh^2\left[\frac{1}{2}\ln\left[\frac{\sqrt{x}+1}{\sqrt{x}-1}\right]\right] = \frac{1}{x},$$

as desired.

3.22: We calculate this proportion by determining the proportion of hyperbolic lines passing through p which intersect ℓ in terms of the angle of their tangent lines at p.

Apply an element of $\mathrm{Möb}(\mathbb{H})$ so that ℓ lies in the Euclidean circle with Euclidean centre 0 and Euclidean radius 1 and so that $p = \lambda i$ for some $\lambda > 1$. Let ℓ_0 be the hyperbolic ray from p to 1, and let φ be the angle between ℓ_0 and the positive imaginary axis I.

We can calculate φ as follows. Note that ℓ_0 lies on the Euclidean circle A with Euclidean centre $c = \frac{1}{2}(1 - \lambda^2)$ and Euclidean radius $r = |c - 1|$, and so the equation of A in $\mathbb{C}$ is $(x-c)^2 + y^2 = r^2$. Differentiating implicitly with respect to x, we get that

$$\frac{\mathrm{d}y}{\mathrm{d}x} = \frac{-x+c}{y}.$$

Hence, at $\lambda i = (0, \lambda)$, we have that the slope of the tangent line to A at $(0, \lambda)$ is

$$\frac{\mathrm{d}y}{\mathrm{d}x}(0, \lambda) = \frac{c}{\lambda} = \frac{1-\lambda^2}{2\lambda}.$$

Hence,

$$\varphi = \arctan\left(\frac{2\lambda}{\lambda^2 - 1}\right),$$

and so the proportion of hyperbolic rays from $p = \lambda i$ intersecting ℓ is

$$\frac{2\varphi}{2\pi} = \frac{1}{\pi}\arctan\left(\frac{2\lambda}{\lambda^2 - 1}\right).$$

Solutions to Chapter 4 exercises:

4.1: Since m^{-1} takes $\mathbb{D}$ to $\mathbb{H}$, we have that

$$\mathrm{length}_{\mathbb{D}}(m \circ f) = \mathrm{length}_{\mathbb{H}}(m^{-1} \circ m \circ f) = \mathrm{length}_{\mathbb{H}}(f),$$

as desired.

4.2: Parametrize the hyperbolic line segment between 0 and r by the path $f : [0, r] \to \mathbb{D}$ given by $f(t) = t$. Since the image of f is the hyperbolic line segment in $\mathbb{D}$ joining 0 and r, we have that $\mathrm{d}_{\mathbb{D}}(0,r) = \mathrm{length}_{\mathbb{D}}(f)$. We have already calculated that

$$\mathrm{d}_{\mathbb{D}}(0,r) = \mathrm{length}_{\mathbb{D}}(f) = \ln\left[\frac{1+r}{1-r}\right].$$

Solving for r as a function of $\mathrm{d}_{\mathbb{D}}(0,r)$, we get that

$$r = \tanh\left[\frac{1}{2}\mathrm{d}_{\mathbb{D}}(0,r)\right].$$

4.3: Apply an element of Möb($\mathbb{D}$) so that ℓ_1 lies in the real axis and so that the point of intersection of ℓ_1 and ℓ_2 is 0.

The endpoints at infinity of ℓ_1 are then $z_1 = 1$ and $z_2 = -1$, and the endpoints at infinity of ℓ_2 are $w_1 = e^{i\theta}$ and $w_2 = -e^{i\theta}$, where θ is the angle between ℓ_1 and ℓ_2.

Hence,

$$[z_1, w_1; z_2, w_2] = [1, e^{i\theta}; -1, -e^{i\theta}] = \frac{-\sin^2(\theta)}{(1-\cos(\theta))^2} = -\cot^2\left(\frac{\theta}{2}\right),$$

and so

$$[z_1, w_1; z_2, w_2]\tan^2\left(\frac{\theta}{2}\right) = -1,$$

as desired.

4.4: Since the hyperbolic radius of S_s is s, the Euclidean radius of S_s is $r = \tanh(\frac{1}{2}s)$, by Exercise 4.2. Parametrize S_s by $f : [0, 2\pi] \to \mathbb{D}$, where $f(t) = r\exp(it)$. Then,

$$\mathrm{length}(f) = \int_f \frac{2}{1-|z|^2}\,|\mathrm{d}z| = \int_0^{2\pi} \frac{2}{1-r^2}\, r\,\mathrm{d}t = \frac{4\pi r}{1-r^2} = 2\pi\sinh(s).$$

4.5: Let $f : [a,b] \to \mathbb{H}$ be a piecewise differentiable path and write $f(t) = x(t) + iy(t)$. Then, $\xi \circ f(t) = x(t) + \frac{1}{2}y(t)i$. Hence,

$$\mathrm{Im}((\xi \circ f)(t)) = \frac{1}{2}y(t)$$

and

$$|(\xi \circ f)'(t)| = \sqrt{(x'(t))^2 + \frac{1}{4}(y'(t))^2},$$

and so

$$\mathrm{length}_X(f) = \int_f \mathrm{d}s_X = \int_a^b \frac{2}{y(t)}\sqrt{(x'(t))^2 + \frac{1}{4}(y'(t))^2}\,\mathrm{d}t.$$

4.6: Since $\operatorname{Im}(\xi(z)) = \operatorname{Im}(iz) = \operatorname{Re}(z)$ and $|\xi'(z)| = |i| = 1$, we see that

$$\mathrm{d}s_X = \frac{1}{\operatorname{Im}(z))}|\xi'(z)||\mathrm{d}z| = \frac{1}{\operatorname{Re}(z)}|\mathrm{d}z|.$$

4.8: Calculating, we see that the curvature is identically -1.

4.9: Calculating, we see that the curvature at $z \in \mathbb{C}$ is identically 16.

Solutions to Chapter 5 exercises:

5.1: Let z_0 and z_1 be two points of $X = \cap_{\alpha\in A} X_\alpha$, and let $\ell_{z_0z_1}$ be the hyperbolic line segment joining z_0 to z_1.

Since each X_α is convex, we have that $\ell_{z_0z_1}$ is contained in each X_α, and so $\ell_{z_0z_1}$ is then contained in their intersection $X = \cap_{\alpha\in A} X_\alpha$.

5.2: The Euclidean radius of D_s is $r = \tanh(\frac{1}{2}s)$. For each θ, let ℓ_θ be the hyperbolic line contained in the Euclidean circle with Euclidean centre on the Euclidean line $\{t\,e^{i\theta} \mid t > 0\}$ and passing through $r\,e^{i\theta}$. Let H_θ be the closed half-plane determined by ℓ_θ containing 0. Since we may express D_s as the intersection $D_s = \cap_\theta H_\theta$ and since each H_θ is convex, we see that D_s is convex.

Since any open hyperbolic disc can be taken by an element of $\text{Möb}(\mathbb{D})$ to some D_s for some $s > 0$ and since $\text{Möb}(\mathbb{D})$ preserves convexity, we see that all open hyperbolic discs are convex.

Repeating this argument with a closed hyperbolic disc and open half-planes H_θ, we see that all closed hyperbolic discs are convex as well.

5.3: By definition, $X \subset \operatorname{conv}(X)$. Conversely, since X is a convex set in the hyperbolic plane containing X, we have that $\operatorname{conv}(X)$ is the intersection of X and other sets, and so $\operatorname{conv}(X) \subset X$. Hence, $X = \operatorname{conv}(X)$.

5.4: Let ℓ_1 and ℓ_2 be the two hyperbolic lines, and let the endpoints at infinity of ℓ_k be x_k and y_k. Set $Z = \{x_1, y_1, x_2, y_2\}$. The convex hull $\operatorname{conv}(\ell_1 \cup \ell_2)$ of the union $\ell_1 \cup \ell_2$ is equal to the convex hull $\operatorname{conv}(Z)$ of Z. This is the region in the plane bounded by four of the six hyperbolic lines determined by these four points.

Note that in the degenerate case that ℓ_1 and ℓ_2 share an endpoint at infinity, the convex hull $\operatorname{conv}(\ell_1 \cup \ell_2)$ is the region bounded by the three hyperbolic lines determined by the three points in Z.

5.5: For example, let ℓ_1 and ℓ_2 be two distinct hyperbolic lines that intersect at some point x. Their union is not convex, since by Exercise 5.4 $\ell_1 \cup \ell_2$ sits as a proper subset of its convex hull. However, $\ell_1 \cup \ell_2$ is star-like with respect to the point of intersection x.

5.6: Let ℓ_{xy} be the closed hyperbolic line segment joining x to y, and let ℓ be the hyperbolic line containing ℓ_{xy}. We can express ℓ as the intersection $\ell = \cap_{\alpha \in A} H_\alpha$ of a collection $\{H_\alpha\}_{\alpha \in A}$ of (two) closed half-planes.

Now let ℓ_x be any hyperbolic line passing through x other than ℓ, and let H_x be the closed half-plane determined by ℓ_x that contains ℓ_{xy}. Similarly, take ℓ_y to be any hyperbolic line other than ℓ passing through y and let H_y be the closed half-plane determined by ℓ_y that contains ℓ_{xy}.

Then, we may express ℓ_{xy} as the intersection

$$\ell_{xy} = H_x \cap H_y \cap \ell = H_x \cap H_y \cap (\cap_{\alpha \in A} H_\alpha)$$

of a collection of closed half-planes.

Since each closed half-plane can be expressed as the intersection of a collection of open half-planes, we can also express ℓ_{xy} as the intersection of a collection of open half-planes.

Now, let ℓ_{xz} be the hyperbolic ray determined by $x \in \mathbb{H}$ and $z \in \overline{\mathbb{R}}$. Let ℓ and ℓ_x be as defined above, and note that

$$\ell_{xz} = H_x \cap \ell = H_x \cap (\cap_{\alpha \in A} H_\alpha).$$

Again, since each closed half-plane can be expressed as the intersection of a collection of open half-planes, we can express ℓ_{xz} as the intersection of a collection of open half-planes.

5.7: Let $\mathcal{H} = \{H_\alpha\}_{\alpha \in A}$ be an uncountable collection of half-planes. Let ℓ_α be the bounding line for H_α. We work in the upper half-plane $\mathbb{H}$ for the sake of concreteness. Let $\mathbb{Q}^+ = \mathbb{Q} \cap (0, \infty)$ denote the set of positive rational numbers.

For each $q \in \mathbb{Q}^+$, consider the hyperbolic disc $U_q(i)$ with hyperbolic centre i and hyperbolic radius q. Since the union $\cup_{q \in \mathbb{Q}^+} U_q(i)$ is equal to $\mathbb{H}$, there is some $q \in \mathbb{Q}^+$ so that $U = U_q(i)$ intersects infinitely many of the bounding lines ℓ_α.

In particular, there is a sequence $\{\ell_{\alpha_n}\}$ of bounding lines, each of that intersects U. For each n, choose a point $x_n \in U \cap \ell_{\alpha_n}$. Since the closure $\overline{U}$ of U is closed and bounded, it is compact, and so there exists a subsequence of $\{x_{\alpha_n}\}$, which we again call $\{x_{\alpha_n}\}$ to avoid proliferation of subscripts, so that $\{x_{\alpha_n}\}$ converges to some point x of U.

By the definition of convergence, for each $\varepsilon > 0$ the hyperbolic disc $U_\varepsilon(x)$ contains infinitely many of the x_{α_n}. Hence, for each $\varepsilon > 0$ the hyperbolic disc $U_\varepsilon(x)$ intersects infinitely many of the bounding lines ℓ_α, and so $\{H_\alpha\}_{\alpha \in A}$ is not locally finite.

5.8: Let P be a hyperbolic polygon, and suppose that P contains three points x, y, and z that do not lie on the same hyperbolic line. Given two points p and q in the hyperbolic plane, let ℓ_{pq} be the closed hyperbolic line segment joining them. Then, the set

$$X = \cup\{\ell_{zp} \mid p \in \ell_{xy}\}$$

has non-empty interior.

In fact, let p be the midpoint of ℓ_{xy} and let q be the midpoint of ℓ_{zp}. Then, the three numbers $\mathrm{d}_{\mathbb{H}}(q, \ell_{xy})$, $\mathrm{d}_{\mathbb{H}}(q, \ell_{xz})$, and $\mathrm{d}_{\mathbb{H}}(q, \ell_{yz})$ are all positive. If we set

$$\varepsilon = \min\{\mathrm{d}_{\mathbb{H}}(q, \ell_{xy}),\ \mathrm{d}_{\mathbb{H}}(q, \ell_{xz}),\ \mathrm{d}_{\mathbb{H}}(q, \ell_{yz})\},$$

then $U_\varepsilon(q)$ is contained in X.

Hence, the only degenerate hyperbolic polygons are closed convex subsets of hyperbolic lines, which are exactly the hyperbolic lines, hyperbolic rays, and hyperbolic line segments.

5.9: Since P has only finitely many sides, and since each side contains exactly two vertices, as it is a hyperbolic line segment, we see that P has exactly as many vertices as it has sides. Let $v_1, \ldots, v_n$ be the vertices of P, and let $V = \{v_1, \ldots, v_n\}$.

By definition, P is a convex set containing V, and so $\mathrm{conv}(V) \subset P$.

Conversely, note that since $\mathrm{conv}(V)$ contains the vertices of P, we have that $\mathrm{conv}(V)$ contains all the sides of P, since each side of P is the hyperbolic line segment joining two of the vertices. That is, we have just shown that ∂P is contained in $\mathrm{conv}(V)$.

Now let x be any point in the interior of P, and let ℓ be any hyperbolic line through x. The intersection of P with ℓ is a hyperbolic line segment ℓ_0 in ℓ whose endpoints are in ∂P.

Hence, since $\mathrm{conv}(V)$ is convex and contains the endpoints of ℓ_0, we have that $\mathrm{conv}(V)$ contains ℓ_0. In particular, x is a point of $\mathrm{conv}(V)$, and so $P \subset \mathrm{conv}(V)$. Hence, $\mathrm{conv}(V) = P$.

5.10: For notational ease, let ℓ_{jk} be the hyperbolic line passing through x_j and x_k. Note that, no matter the value of s, the hyperbolic lines ℓ_{12} and ℓ_{34} are parallel, since they are contained in parallel Euclidean lines.

The hyperbolic line ℓ_{13} is contained in the Euclidean circle C_{13} with Euclidean centre 0 and Euclidean radius $\sqrt{2}$, while the hyperbolic line is contained in the Euclidean circle C_{24} with Euclidean centre $\frac{1}{4}s^2 - 1$ and Euclidean radius $\frac{1}{4}\sqrt{s^4 + 64}$. Note that C_{13} and C_{24} intersect at a point of $\mathbb{R}$ exactly when $\pm\sqrt{2}$ lies on C_{24}.

Calculating, we have that $-\sqrt{2}$ lies on C_{24} precisely when

$$\left|\left(\frac{1}{4}s^2 - 1\right) + \sqrt{2}\right| = \frac{1}{4}\sqrt{s^4 + 64},$$

namely

$$s = \sqrt{10 + 6\sqrt{2}}.$$

Similarly, we have that $\sqrt{2}$ lies on C_{24} precisely when

$$\left|\left(\frac{1}{4}s^2 - 1\right) - \sqrt{2}\right| = \frac{1}{4}\sqrt{s^4 + 64},$$

namely

$$s = \sqrt{10 - 6\sqrt{2}}.$$

Hence, Q_s is a hyperbolic parallelogram if and only if

$$\sqrt{10 - 6\sqrt{2}} \leq s \leq \sqrt{10 + 6\sqrt{2}}.$$

5.11: This is the same as the proof of Exercise 5.9.

5.12: We first note that every hyperbolic triangle is contained in an ideal hyperbolic triangle. To see this, let T be a hyperbolic triangle with vertices v_1, v_2, and v_3, and let x be any point in the interior of T.

Let y_k to be the endpoint at infinity of the hyperbolic ray determined by x and v_k, and let P be the ideal triangle with ideal vertices y_1, y_2, and y_3. Then, P contains T. Hence, it suffices to work in the case that T is an ideal triangle.

To make the calculation easier, let m be an element of Möb($\mathbb{H}$) taking T to the ideal triangle with ideal vertices at $y_1 = 0$, $y_2 = 1$, and $y_3 = \infty$. Let ℓ_{jk} be the hyperbolic line determined by y_j and y_k.

For each $r > 0$, let C_r be the Euclidean circle with Euclidean centre 0 and Euclidean radius r. Note that C_r intersects ℓ_{12} at the point $re^{i\theta}$, where $\cos(\theta) = r$. (This relationship between r and θ is obtained by noticing that $re^{i\theta}$ also lies in the Euclidean circle $\left(x - \frac{1}{2}\right)^2 + y^2 = \frac{1}{4}$, which contains the hyperbolic line ℓ_{12}.) For each point ri on ℓ_{13}, the hyperbolic distance between ri and ℓ_{12} is equal to the hyperbolic distance between ri and $re^{i\theta}$

By the solution to Exercise 3.19, the hyperbolic distance between ri and $re^{i\theta}$ is

$$d_{\mathbb{H}}(re^{i\theta}, ri) = \ln\left[\frac{1 + \cos(\theta)}{\sin(\theta)}\right].$$

By symmetry, we need only consider θ in the range $[\frac{\pi}{4}, \frac{\pi}{2}]$. On $[\frac{\pi}{4}, \frac{\pi}{2}]$, the function $\ln\left[\frac{1+\cos(\theta)}{\sin(\theta)}\right]$ is decreasing, and so $d_{\mathbb{H}}(re^{i\theta}, ri)$ is maximized at $\theta = \frac{\pi}{4}$. Hence,

$$d_{\mathbb{H}}(x, \ell_{12}) \leq \ln\left[\frac{1+\frac{1}{\sqrt{2}}}{\frac{1}{\sqrt{2}}}\right] = \ln(\sqrt{2}+1).$$

5.13: The hyperbolic area of X_s is

$$\text{area}_{\mathbb{H}}(X_s) = \int_{X_s} \frac{1}{y^2}\, dx\, dy = \int_{-1}^{1}\int_{s}^{\infty} \frac{1}{y^2}\, dy\, dx = \frac{2}{s}.$$

5.14: Rewriting in terms of x and y, we see that $B(x,y) = (-x, y)$. Hence,

$$DB(x,y) = \begin{pmatrix} -1 & 0 \\ 0 & 1 \end{pmatrix},$$

and so

$$\det(DB(x,y)) = -1.$$

Since

$$h \circ B(x,y) = \frac{1}{y^2},$$

the change of variables theorem yields that

$$\text{area}_{\mathbb{H}}(B(X)) = \int_{B(X)} \frac{1}{y^2}\, dx\, dy = \int_X \frac{1}{y^2}\, dx\, dy = \text{area}_{\mathbb{H}}(X),$$

as desired.

5.15: Rewriting in terms of x and y, we see that $f(x,y) = (x+y, y)$. Hence,

$$Df(x,y) = \begin{pmatrix} 1 & 1 \\ 0 & 1 \end{pmatrix},$$

and so

$$\det(Df(x,y)) = 1.$$

Since

$$h \circ f(x,y) = \frac{1}{y^2},$$

the change of variables theorem yields that

$$\text{area}_{\mathbb{H}}(f(X)) = \int_{f(X)} \frac{1}{y^2}\, dx\, dy = \int_X \frac{1}{y^2}\, dx\, dy = \text{area}_{\mathbb{H}}(X).$$

This completes the proof that f preserves hyperbolic area.

To see that f is not an element of $\text{Möb}(\mathbb{H})$, note that f takes the Euclidean line $\{\text{Re}(z) = 1\}$, which contains a hyperbolic line, to the Euclidean line $\{\text{Re}(z) =$

$1+\text{Im}(z)\}$, which does not intersect $\mathbb{R}$ perpendicularly and so does not contain a hyperbolic line.

5.16: Let s_{jk} be the side of P joining the vertex v_j of P to the vertex v_k of P, let ℓ_{jk} be the hyperbolic line containing s_{jk}, and let C_{jk} be the Euclidean circle containing ℓ_{jk}.

The equation of C_{12} is $\left|z-\frac{7}{4}\right| = \frac{\sqrt{65}}{4}$. The equation of C_{23} is $\left|z-\frac{9}{4}\right| = \frac{\sqrt{65}}{4}$. The equation of C_{13} is $|z-2| = \sqrt{5}$. So, the hyperbolic area of P is given by the integral

$$\begin{aligned} \text{area}_{\mathbb{H}}(P) &= \int_P \frac{1}{y^2}\,\mathrm{d}x\,\mathrm{d}y \\ &= \int_0^2 \int_{\sqrt{\frac{65}{16}-(x-\frac{7}{4})^2}}^{\sqrt{5-(x-2)^2}} \frac{1}{y^2}\,\mathrm{d}y\,\mathrm{d}x + \int_2^4 \int_{\sqrt{\frac{65}{16}-(x-\frac{9}{4})^2}}^{\sqrt{5-(x-2)^2}} \frac{1}{y^2}\,\mathrm{d}y\,\mathrm{d}x \end{aligned}$$

5.17: We begin with the fact that if C_1 and C_2 are intersecting Euclidean circles, where C_k has Euclidean centre c_k and Euclidean radius r_k, then using the law of cosines, the angle θ between C_1 and C_2 satisfies

$$|c_1-c_2|^2 = r_1^2 + r_2^2 - 2r_1r_2\cos(\theta),$$

and so

$$\cos(\theta) = \frac{r_1^2 + r_2^2 - |c_1-c_2|^2}{2r_1r_2}.$$

By the solution to Exercise 5.16, we have that C_{12} has Euclidean centre $\frac{7}{4}$ and Euclidean radius $\frac{\sqrt{65}}{4}$; that C_{23} has Euclidean centre $\frac{9}{4}$ and Euclidean radius $\frac{\sqrt{65}}{4}$; and that C_{13} has Euclidean centre 2 and Euclidean radius $\sqrt{5}$.

The angle α between C_{12} and C_{13} is given by

$$\cos(\alpha) = \frac{\frac{65}{16}+5-|\frac{7}{4}-2|^2}{2\frac{\sqrt{65}}{4}\sqrt{5}} = \frac{18}{\sqrt{325}},$$

namely

$$\alpha \sim 0.0555.$$

The angle β between C_{23} and C_{13} is given by

$$\cos(\beta) = \frac{\frac{65}{16}+5-|\frac{9}{4}-2|^2}{2\frac{\sqrt{65}}{4}\sqrt{5}} = \frac{18}{\sqrt{325}},$$

namely

$$\beta \sim 0.0555.$$

The angle γ between C_{12} and C_{23} is given by

$$\cos(\gamma) = \frac{\frac{65}{16} + \frac{65}{16} - |\frac{7}{4} - \frac{9}{4}|^2}{2\frac{\sqrt{65}}{4}\frac{\sqrt{65}}{4}} = \frac{126}{130},$$

namely

$$\gamma \sim 0.2487.$$

Hence, we see by Theorem 5.15 that

$$\text{area}_{\mathbb{H}}(P) = \pi - (\alpha + \beta + \gamma) \sim 2.7819.$$

5.18: The hyperbolic radius s of D_s is related to the Euclidean radius R by $R = \tanh(\frac{1}{2}s)$. The hyperbolic area of D_s is then

$$\begin{aligned} \text{area}_{\mathbb{D}}(D_s) &= \int_{D_s} \frac{4r}{(1-r^2)^2} \, \mathrm{d}r \, \mathrm{d}\theta \\ &= \int_0^R \int_0^{2\pi} \frac{4r}{(1-r^2)^2} \, \mathrm{d}r \, \mathrm{d}\theta \\ &= 2\pi \int_0^R \frac{4r}{(1-r^2)^2} \, \mathrm{d}r \, \mathrm{d}\theta = \frac{4\pi R^2}{1-R^2} = 4\pi \sinh^2\left(\frac{1}{2}s\right). \end{aligned}$$

5.19: Since $\text{length}_{\mathbb{D}}(S_s) = 2\pi \sinh(s)$ and since $\text{area}_{\mathbb{D}}(D_s) = 4\pi \sinh^2\left(\frac{1}{2}s\right)$, we have that

$$q_{\mathbb{D}}(s) = \frac{\text{length}_{\mathbb{D}}(S_s)}{\text{area}_{\mathbb{D}}(D_s)} = \frac{2\pi \sinh(s)}{4\pi \sinh^2\left(\frac{1}{2}s\right)} = \coth\left(\frac{1}{2}s\right).$$

In particular, note that $q_{\mathbb{D}}(s) \to 1$ as $s \to \infty$, and $q_{\mathbb{D}}(s) \to \infty$ as $s \to 0$.

The corresponding Euclidean quantity $q_{\mathbb{C}}(r) = \frac{2}{r}$ behaves much differently as the radius of the Euclidean circle and Euclidean disc get large. Namely, $q_{\mathbb{C}}(r) \to 0$ as $r \to \infty$, while we again have that $q_{\mathbb{C}}(r) \to \infty$ as $r \to 0$.

5.20: Let C_0 be the Euclidean circle in $\mathbb{C}$ containing the hyperbolic line ℓ_0 passing through $rp_0 = r$ and $rp_1 = r\exp\left(\frac{2\pi i}{n}\right)$. The Euclidean centre of C_0 is then of the form $s\exp\left(\frac{\pi i}{n}\right)$ for some $s > 1$. Since C_0 must intersect $\mathbb{S}^1$ perpendicularly, we have from Exercise 1.2 that the Euclidean radius of C_0 is $\sqrt{s^2-1}$.

In order for C_0 to pass through r, we must have that

$$\left| s\exp\left(\frac{\pi i}{n}\right) - r \right| = \sqrt{s^2-1},$$

and so

$$s = \frac{r^2+1}{2r\cos\left(\frac{\pi}{n}\right)}.$$

In particular, the Euclidean centre of C_0 is

$$s \exp\left(\frac{\pi i}{n}\right) = \frac{r^2+1}{2r\cos\left(\frac{\pi}{n}\right)} \exp\left(\frac{\pi i}{n}\right)$$

and the Euclidean radius of C_0 is

$$\sqrt{s^2-1} = \sqrt{\frac{(r^2+1)^2}{4r^2\cos^2\left(\frac{\pi}{n}\right)} - 1}.$$

We can repeat this calculation for the Euclidean circle C_{n-1} containing the hyperbolic line ℓ_{n-1} passing through $rp_0 = r$ and $rp_{n-1} = r\exp\left(\frac{2\pi(n-1)i}{n}\right)$. The Euclidean centre of C_{n-1} is

$$s \exp\left(\frac{-\pi i}{n}\right) = \frac{r^2+1}{2r\cos\left(\frac{\pi}{n}\right)} \exp\left(\frac{-\pi i}{n}\right)$$

and the Euclidean radius of C_{n-1} is

$$\sqrt{s^2-1} = \sqrt{\frac{(r^2+1)^2}{4r^2\cos^2\left(\frac{\pi}{n}\right)} - 1}.$$

The interior angle $\alpha(r)$ of $P_n(r)$ at $r = rp_0$ is equal to the angle between C_0 and C_{n-1}, and hence satisfies

$$\begin{aligned}
\cos(\alpha(r)) &= \frac{2(s^2-1) - \left|s\exp\left(\frac{\pi i}{n}\right) - s\exp\left(\frac{-\pi i}{n}\right)\right|^2}{2(s^2-1)} \\
&= \frac{2(s^2-1) - 4s^2\sin^2\left(\frac{\pi}{n}\right)}{2(s^2-1)} \\
&= 1 - \frac{2(r^2+1)^2\sin^2\left(\frac{\pi}{n}\right)}{(r^2+1)^2 - 4r^2\cos^2\left(\frac{\pi}{n}\right)}.
\end{aligned}$$

The continuity of $\alpha(r)$ follows immediately from the continuity of the right hand side of this expression.

5.21: For $n \geq 5$, the interval of possible angles of a regular hyperbolic n-gon is $(0, \frac{n-2}{n}\pi)$. Since $\frac{n-2}{n} > \frac{1}{2}$ for $n \geq 5$, this interval contains $\frac{1}{2}\pi$, and so there exists a regular hyperbolic n-gon with all right angles.

5.22: The hyperbolic length of the side of $P_n(r)$ joining $rp_0 = r$ to $rp_1 = r\exp\left(\frac{2\pi i}{n}\right)$ is equal to $d_{\mathbb{D}}\left(r, r\exp\left(\frac{2\pi i}{n}\right)\right)$.

Set $\theta = \frac{2\pi}{n}$. To calculate $d_{\mathbb{D}}(r, re^{i\theta})$, we first choose an element m of $\text{Möb}(\mathbb{D})$ satisfying $m(r) = 0$. Write $m(z) = \frac{\alpha z+\beta}{\overline{\beta}z+\overline{\alpha}}$, where $|\alpha|^2 - |\beta|^2 = 1$.

Since $m(r) = \frac{\alpha r+\beta}{\overline{\beta}r+\overline{\alpha}} = 0$, we have that $\beta = -\alpha r$, and so

$$m(z) = \frac{\alpha z - \alpha r}{-\overline{\alpha}rz + \overline{\alpha}} = \frac{\alpha(z-r)}{\overline{\alpha}(-rz+1)}.$$

Thus,

$$m(re^{i\theta}) = \frac{\alpha r(e^{i\theta}-1)}{\overline{\alpha}(-r^2e^{i\theta}+1)}$$

and

$$|m(re^{i\theta})| = \left|\frac{\alpha r(e^{i\theta}-1)}{\overline{\alpha}(-r^2e^{i\theta}+1)}\right| = \left|\frac{r(e^{i\theta}-1)}{-r^2e^{i\theta}+1}\right|.$$

Hence,

$$\begin{aligned}
&\mathrm{d}_{\mathbb{D}}(r, re^{i\theta})\\
&\quad= \mathrm{d}_{\mathbb{D}}(m(r), m(re^{i\theta}))\\
&\quad= \mathrm{d}_{\mathbb{D}}(0, m(re^{i\theta}))\\
&\quad= \ln\left[\frac{1+|m(re^{i\theta})|}{1-|m(re^{i\theta})|}\right]\\
&\quad= \ln\left[\frac{|-r^2e^{i\theta}+1|+|r(e^{i\theta}-1)|}{|-r^2e^{i\theta}+1|-|r(e^{i\theta}-1)|}\right]\\
&\quad= \ln\left[\frac{(1+r^2)^2-4r^2\cos(\theta)+2r\sqrt{2(1-2r^2\cos(\theta)+r^4)(1-\cos(\theta))}}{(1-r^2)^2}\right].
\end{aligned}$$

5.23: These follow directly, with some algebraic massage, from the definition of $\cosh(x)$ and $\sinh(x)$ in terms of e^x, namely

$$\cosh(x) = \frac{1}{2}\left(e^x + e^{-x}\right) \text{ and } \sinh(x) = \frac{1}{2}\left(e^x - e^{-x}\right).$$

5.24: Since both $\sinh(c)$ and $\sin(\gamma)$ are positive, we consider instead the quantity $\frac{\sinh^2(c)}{\sin^2(\gamma)}$.

Write $A = \cosh(a)$, $B = \cosh(b)$, and $C = \cosh(c)$. By the law of cosines I, we have that

$$\sin^2(\gamma) = 1 - \cos^2(\gamma) = 1 - \left(\frac{AB - C}{\sinh(a)\sinh(b)}\right)^2.$$

Multiplying through, we get

$$\begin{aligned}
\sin^2(\gamma)\sinh^2(a)\sinh^2(b) &= \sinh^2(a)\sinh^2(b) - (AB-C)^2\\
&= \sinh^2(a)\sinh^2(b) - A^2B^2 - C^2 + 2ABC\\
&= (A^2-1)(B^2-1) - A^2B^2 - C^2 + 2ABC\\
&= A^2B^2 - A^2 - B^2 + 1 - A^2B^2 - C^2 + 2ABC\\
&= 1 - A^2 - B^2 - C^2 + 2ABC.
\end{aligned}$$

Hence, we have that

$$\frac{\sin^2(\gamma)}{\sinh^2(c)} = \frac{1 - A^2 - B^2 - C^2 + 2ABC}{\sinh^2(a)\sinh^2(b)\sinh^2(c)}.$$

Since the right hand side remains unchanged after permuting a, b, and c, and simultaneously permuting α, β, and γ, the left hand side must be unchanged as well, and so we see that

$$\frac{\sinh^2(c)}{\sin^2(\gamma)} = \frac{\sinh^2(b)}{\sin^2(\beta)} = \frac{\sinh^2(a)}{\sin^2(\alpha)}.$$

Taking square roots, we obtain the law of sines.

For the law of cosines II, start with the law of cosines I, which gives that

$$\cos(\gamma) = \frac{AB - C}{\sinh(a)\sinh(b)} = \frac{AB - C}{\sqrt{(A^2-1)(B^2-1)}}.$$

Applying the law of cosines I to the other two vertices gives

$$\cos(\alpha) = \frac{BC - A}{\sqrt{(B^2-1)(C^2-1)}} \text{ and } \sin(\alpha) = \frac{\sqrt{1 + 2ABC - A^2 - B^2 - C^2}}{\sqrt{(B^2-1)(C^2-1))}},$$

and that

$$\cos(\beta) = \frac{AC - B}{\sqrt{(A^2-1)(C^2-1)}} \text{ and } \sin(\beta) = \frac{\sqrt{1 + 2ABC - A^2 - B^2 - C^2}}{\sqrt{(A^2-1)(C^2-1))}}.$$

Hence,

$$\begin{aligned} \frac{\cos(\gamma) + \cos(\alpha)\cos(\beta)}{\sin(\alpha)\sin(\beta)} &= \frac{(BC - A)(AC - B) + (AB - C)(C^2 - 1)}{1 + 2ABC - A^2 - B^2 - C^2} \\ &= C = \cosh(c), \end{aligned}$$

as desired.

5.25: Consider the hyperbolic law of cosines I with $\alpha = \frac{\pi}{2}$. Let a be the hyperbolic length of the side of T opposite the vertex with angle α, and let b and c be the hyperbolic lengths of the sides adjacent to the vertex with angle α. Then,

$$\cosh(a) = \cosh(b)\cosh(c).$$

5.26: The fact that the three interior angles are equal follows immediately from the hyperbolic law of cosines I, namely that

$$\cos(\alpha) = \frac{\cosh^2(a) - \cosh(a)}{\sinh^2(a)}.$$

Consider the hyperbolic triangle T' formed by bisecting the hyperbolic triangle T, so that T' has angles α, $\frac{1}{2}\alpha$, and $\frac{1}{2}\pi$, and has the corresponding hyperbolic lengths of the opposite sides being b, $\frac{1}{2}a$, and a, where b is as yet undetermined.

Applying the hyperbolic law of cosines I to T', we obtain

$$\begin{aligned}\cos\left(\frac{1}{2}\alpha\right) &= -\cos\left(\frac{1}{2}\pi\right)\cos(\alpha)+\sin\left(\frac{1}{2}\pi\right)\sin(\alpha)\cosh\left(\frac{1}{2}a\right)\\ &= \sin(\alpha)\cosh\left(\frac{1}{2}a\right)\\ &= 2\sin\left(\frac{1}{2}\alpha\right)\cos\left(\frac{1}{2}\alpha\right)\cosh\left(\frac{1}{2}a\right).\end{aligned}$$

Dividing through by $\cos\left(\frac{1}{2}\alpha\right)$, we obtain

$$1 = 2\sin\left(\frac{1}{2}\alpha\right)\cosh\left(\frac{1}{2}a\right),$$

as desired.

Solutions to Chapter 6 exercises:

6.1: Every point a on A has the form $a = \alpha e^{i\theta}$. Hence, $m(a) = \lambda a = \lambda\alpha e^{i\theta}$.

Calculating, we see that

$$\begin{aligned}\mathrm{d}_{\mathbb{H}}\left(\alpha e^{i\theta}, \lambda\alpha e^{i\theta}\right) &= \mathrm{d}_{\mathbb{H}}\left(e^{i\theta}, \lambda e^{i\theta}\right)\\ &= \mathrm{d}_{\mathbb{H}}\left(e^{i\theta}-\cos(\theta), \lambda e^{i\theta}-\cos(\theta)\right)\\ &= \mathrm{d}_{\mathbb{H}}\left(i\sin(\theta), (\lambda-1)\cos(\theta)+i\lambda\sin(\theta)\right)\\ &= \mathrm{d}_{\mathbb{H}}\left(i, (\lambda-1)\cot(\theta)+\lambda i\right).\end{aligned}$$

Write $(\lambda-1)\cot(\theta)+\lambda i = \rho^{i\varphi}$. Calculating, we see that

$$\rho = \sqrt{(\lambda-1)^2\cot^2(\theta)+\lambda^2}$$

and

$$\csc(\varphi) = \frac{\rho}{\lambda}.$$

By Exercise 5.25, we have that $\mathrm{d}_{\mathbb{H}}\left(i, (\lambda-1)\cot(\theta)+i\lambda\right)$ satisfies

$$\cosh(\mathrm{d}_{\mathbb{H}}(i, (\lambda-1)\cot(\theta)+\lambda i)) = \cosh(\mathrm{d}_{\mathbb{H}}(i,\rho i))\cosh(\mathrm{d}_{\mathbb{H}}(\rho i, \rho e^{i\varphi})).$$

By the solution to Exercise 3.19 we have that

$$\mathrm{d}_{\mathbb{H}}(\rho i, \rho e^{i\varphi}) = \ln\left[\frac{1+\cos(\varphi)}{\sin(\varphi)}\right],$$

and so

$$\cosh(\mathrm{d}_{\mathbb{H}}(\rho i, \rho e^{i\varphi})) = \csc(\varphi).$$

Hence,

$$\begin{aligned}\cosh\left(\mathrm{d}_{\mathbb{H}}\left(\alpha e^{i\theta}, \lambda\alpha e^{i\theta}\right)\right) &= \cosh(\ln(\rho))\csc(\varphi) \\ &= \frac{1}{2\lambda}(\rho^2+1) \\ &= \frac{1}{2}\left(\frac{(\lambda-1)^2}{\lambda}\cot^2(\theta) + \frac{\lambda^2+1}{\lambda}\right) \\ &= (\cosh(\ln(\lambda)) - 1)\cot^2(\theta) + \cosh(\ln(\lambda)).\end{aligned}$$

6.2: Since the hyperbolic line ℓ determined by $\sqrt{2}$ and $-\sqrt{2}$ lies in the Euclidean circle with Euclidean centre 0 and Euclidean radius $\sqrt{2}$, reflection in ℓ is given by $r(z) = \frac{2}{\overline{z}}$.

Composing, we see that

$$r \circ q(z) = \frac{2z+2}{z+2} = \frac{\sqrt{2}z + \sqrt{2}}{\frac{1}{\sqrt{2}}z + \sqrt{2}} = m(z),$$

which is loxodromic fixing $\sqrt{2}$ and $-\sqrt{2}$.

Since $r(z) = r^{-1}(z)$, we then have that $q(z) = r \circ m(z)$, as desired.

6.3: Setting $q(z) = z$, we get $z = -\overline{z} + 1$, which we can rewrite as $\mathrm{Re}(z) = \frac{1}{2}$. Hence, the fixed points of q are exactly the points on the hyperbolic line in $\mathbb{H}$ contained in the Euclidean line $\{\mathrm{Re}(z) = \frac{1}{2}\}$.

6.4: The fixed points in $\overline{\mathbb{C}}$ of

$$q(z) = \frac{2i\overline{z} - i}{3i\overline{z} - 2i}$$

are the solutions in $\overline{\mathbb{C}}$ of $q(z) = z$, which are the points z in $\overline{\mathbb{C}}$ satisfying

$$3i|z|^2 - 2i(z + \overline{z}) + i = 0.$$

Writing $z = x + iy$, we see that the fixed points of q in $\overline{\mathbb{C}}$ are the points on the Euclidean circle

$$x^2 - \frac{4}{3}x + \frac{1}{3} + y^2 = 0,$$

which is the Euclidean circle

$$\left(x - \frac{2}{3}\right)^2 + y^2 = \frac{1}{9}.$$

Hence, the fixed points of q are exactly the points on the hyperbolic line in $\mathbb{H}$ contained in the Euclidean circle with Euclidean centre $\frac{2}{3}$ and Euclidean radius $\frac{1}{3}$.

6.5: There are two cases. Suppose that n is reflection in the line $\{x = a\}$. From our work in Section 2.6, we can write $n(z) = -\overline{z} + 2a$ for some real number a. Calculating, we see that

$$p \circ n(z) = p(-\overline{z} + 2a) = -\overline{z} + 1 - 2a.$$

In fact, $p \circ n$ fixes every point z in $\mathbb{H}$ with $2\,\mathrm{Re}(z) = 1 - 2a$, and so $p \circ n$ is reflection in the hyperbolic line $\{x = \frac{1}{2}(1 - 2a)\}$ in $\mathbb{H}$.

Suppose now that n is reflection in the line contained in the Euclidean circle with Euclidean centre c and Euclidean radius r. By our work in Section 2.6, we can write $n(z) = c + \frac{r^2}{\overline{z}-c}$. Calculating, we see that

$$p \circ n(z) = c + 1 + \frac{r^2}{\overline{z} - c} = \frac{(c+1)\overline{z} + r^2 - c(c+1)}{\overline{z} - c}.$$

Setting $p \circ n(z) = z$, we see that the fixed points of $p \circ n$ are those points in $\overline{\mathbb{C}}$ satisfying

$$x^2 - (2c+1)x + y^2 + c(c+1) - r^2 - iy = 0.$$

Setting imaginary parts equal, we see that $y = 0$. In particular, we see that $p \circ n$ has no fixed points in $\mathbb{H}$, and the fixed points in $\overline{\mathbb{R}}$ are the two solutions to

$$\left(x - \frac{1}{2}(2c+1)\right)^2 = r^2 + \frac{1}{4}.$$

6.6: We use a counting argument. Let X be an uncountable subset of $\mathbb{H}$, and assume that X is discrete. Hence, for each $x \in X$ there exists some $\varepsilon(x) > 0$ so that $U_{\varepsilon(x)}(x) \cap X = \{x\}$. Note that we can express the positive reals $(0, \infty)$ as the union of countably many intervals; for example we may write

$$(0, \infty) = [1, \infty) \cup (\cup_{n \in \mathbb{N}} [1/(n+1), 1/n]).$$

Since there are uncountably many points of X and countably many intervals, it must be that uncountably many of the $\varepsilon(x)$ lie in the same interval. Hence, there is some $\varepsilon > 0$ so that

$$Y = \{x \in X \mid \varepsilon(x) \geq \varepsilon\}$$

is uncountable.

Set $\delta = \frac{1}{2}\varepsilon$. By the definition of ε, the hyperbolic discs $U_\delta(x_1)$ and $U_\delta(x_2)$ are disjoint for any pair x_1, x_2 of distinct points of Y.

Since the countable set $Q = \{p+iq | p, q \in \mathbb{Q}\}$ is dense in $\mathbb{H}$, each $U_\varepsilon(x)$ contains an element of Q. Since the $U_\delta(x)$ are disjoint as x ranges over Y, no two $U_\delta(x)$ contain the same element of Q. Since there are uncountably many $U_\delta(x)$, this implies that there are uncountably many elements of Q, which contradicts the countability of Q.

6.7: Let Φ be a subgroup of Γ, and suppose that there exists a non-trivial element $m_\alpha(z) = \alpha z$ of Φ, where $\alpha > 1$. Note that m_α generates an infinite cyclic subgroup of Φ, namely

$$\langle m_\alpha \rangle = \{m_\alpha^n(z) = \alpha^n z \mid n \in \mathbb{Z}\}.$$

Suppose that $m_\beta(z) = \beta z$ is an element of Φ that is not a power of m_α, for some $\beta > 1$. For each k in $\mathbb{N}$, choose n_k so that β^k lies between α^{n_k} and α^{n_k+1}.

Then, for each $k \in \mathbb{N}$, $m_\beta^k \circ m_\alpha^{-n_k}(i)$ lies on the positive imaginary axis I between i and αi. Moreover, since no non-zero power of m_β lies in $\langle m_\alpha \rangle$, we have that the points $\{m_\beta^k \circ m_\alpha^{-n_k}(i)\}$ are distinct.

Since the closed hyperbolic line segment I_α joining i and αi is compact, there is a subsequence of $\{m_\beta^k \circ m_\alpha^{-n_k}(i)\}$ that converges to a point of I_α. As we did in Section 6.2, we can now find a sequence $\{\varphi_n\}$ in Φ so that $\{\varphi(i)\}$ converges to i.

Hence, if Φ is not infinite cyclic, then Φ is not discrete.

6.8: Let Γ be a discrete subgroup of Möb($\mathbb{H}$) and let z be any point of $\mathbb{H}$. By definition, the set $\Gamma(z) = \{\gamma(z) \mid \gamma \in \Gamma\}$ is discrete, and hence is countable by Exercise 6.6.

Since Γ_z is finite, by Theorem 6.6, we have that the stabilizer $\Gamma_{\gamma(z)}$ is finite for every element γ of Γ.

Now we count. There are only countably many points in $\Gamma(z)$, and the stabilizer of each point is finite. Hence, there can be only countably many elements of Γ.

6.9: This particular solution uses the same trick as is used in the proof of Proposition 6.7. Suppose that $\{\gamma_n(z) = \frac{a_n z + b_n}{c_n z + d_n}\}$ is a sequence of distinct elements of Γ converging to an element $\varphi(z) = \frac{az+b}{cz+d}$ of Möb$^+(\mathbb{H})$, where all are normalized to have determinant equal to 1. In particular, we have that $a_n \to a$, $b_n \to b$, $c_n \to c$, and $d_n \to d$ as $n \to \infty$.

Since $\gamma_{n+1}^{-1}(z) = \frac{d_{n+1}z - b_{n+1}}{-c_{n+1}z + a_{n+1}}$, we have that

$$\gamma_{n+1}^{-1} \circ \gamma_n(z) = \frac{(a_n d_{n+1} - b_{n+1} c_n)z + b_n d_{n+1} - b_{n+1} d_n}{(a_{n+1} c_n - a_n c_{n+1})z + a_{n+1} d_n - b_n c_{n+1}},$$

and so we can see directly that $\{\gamma_{n+1}^{-1} \circ \gamma_n\}$ converges to the identity.

If Γ contains a sequence of distinct elements converging to the identity, then obviously Γ contains a sequence of distinct elements converging to an element φ of Möb($\mathbb{H}$), namely take φ to be the identity.

6.10: Suppose that Φ is discrete. Since $\Phi \cap \Theta$ is a subgroup of Φ, Proposition 6.9 implies that $\Phi \cap \Theta$ is discrete. Since $\Phi \cap \Theta$ has finite index in Θ, Proposition 6.11 implies that Θ is discrete. The proof of the other direction, that Θ is discrete implies that Φ is discrete, is the same.

6.11: Suppose that $\overline{U}$ contains a fundamental set for the action of Γ on $\mathbb{H}$. Hence, for each point x of $\mathbb{H}$, there is an element γ_x of Γ so that $\gamma_x(x)$ is contained in $\overline{U}$. Applying γ_x^{-1}, we have that x is contained in $\gamma_x^{-1}(\overline{U})$. Since x is an arbitrary point of $\mathbb{H}$, we have that $\Gamma(\overline{U}) = \mathbb{H}$.

Conversely, suppose that $\Gamma(\overline{U}) = \mathbb{H}$. Then, if y is any point of $\mathbb{H}$, there exists an element γ_y of Γ so that y lies in $\gamma_y(\overline{U})$. Applying γ_y^{-1}, we have that $\gamma_y^{-1}(y)$ is a point of $\overline{U}$. Hence, $\overline{U}$ contains a point of the orbit $\Gamma(y)$ of y under Γ for every point y of $\mathbb{H}$. Hence, $\overline{U}$ contains a fundamental set for the action of Γ on $\mathbb{H}$.

6.12: First, note that the orbit of 0 under Θ is

$$\Theta(0) = \{n + m(1+2i) \mid n, m \in \mathbb{Z}\} = \{p + 2qi \mid p, q \in \mathbb{Z}\}.$$

So, $D_\Theta(0)$ is the hexagonal region in $\mathbb{C}$ bounded by the Euclidean lines $\ell_1 = \{\mathrm{Im}(z) = 1\}$, which is the perpendicular bisector of the line segment joining 0 and $2i$; $\ell_2 = \{\mathrm{Im}(z) = -1\}$, which is the perpendicular bisector of the line segment joining 0 and $-2i$; $\ell_3 = \{\mathrm{Re}(z) = \frac{1}{2}\}$, which is the perpendicular bisector of the line segment joining 0 and 1; and $\ell_4 = \{\mathrm{Re}(z) = -\frac{1}{2}\}$, which is the perpendicular bisector of the line segment joining 0 and -1.

6.13: Let v_k be the vertex adjacent to the sides s_k and s_{k+1} of P, where $s_9 = s_1$. We determine the cycle transformation associated to v_1. v_1 is adjacent to s_1 and s_2, so start with s_1. The side pairing transformation associated to s_1 is γ_1^{-1}, and $\gamma_1^{-1}(v_1) = v_4$.

v_4 is adjacent to $s_5 = \gamma_1^{-1}(s_1)$ and s_4. The side pairing transformation associated to s_4 is γ_4^{-1}, and $\gamma_4^{-1}(v_4) = v_7$.

v_7 is adjacent to $s_8 = \gamma_4^{-1}(s_4)$ and s_7. The side pairing transformation associated to s_7 is γ_3, and $\gamma_3(v_7) = v_2$.

v_2 is adjacent to $s_3 = \gamma_3(s_7)$ and s_2. The side pairing transformation associated to s_2 is γ_2^{-1}, and $\gamma_2^{-1}(v_2) = v_5$.

v_5 is adjacent to $s_6 = \gamma_2^{-1}(s_2)$ and s_5. The side pairing transformation associated to s_5 is γ_1, and $\gamma_1(v_5) = v_8$.

v_8 is adjacent to $s_1 = \gamma_1(s_5)$ and s_8. The side pairing transformation associated to s_8 is γ_4, and $\gamma_4(v_8) = v_3$.

v_3 is adjacent to $s_4 = \gamma_4(s_8)$ and s_3. The side pairing transformation associated to s_3 is γ_3^{-1}, and $\gamma_3^{-1}(v_3) = v_6$.

v_6 is adjacent to $s_7 = \gamma_3^{-1}(s_3)$ and s_6. The side pairing transformation associated to s_6 is γ_2, and $\gamma_2(v_6) = v_1$.

So, the cycle transformation φ associated to v_1 is

$$\varphi = \gamma_2 \circ \gamma_3^{-1} \circ \gamma_4 \circ \gamma_1 \circ \gamma_2^{-1} \circ \gamma_3 \circ \gamma_4^{-1} \circ \gamma_1^{-1},$$

which is the identity and so has order $r_\varphi = 1$.

Since each vertex appears exactly once in the vertex cycle that starts with v_1, we have that the angle sum of v_1 is $\text{sum}(v_1) = 8\,\frac{1}{4}\pi = 2\pi$. So, we have that $r_\varphi\,\text{sum}(v_1) = 2\pi$, as desired.

If we calculate the vertex cycle and cycle transformation starting with any other vertex, we get a permutation of the vertex cycle for v_1 and a cycle transformation that is a conjugate, or the inverse of a conjugate, of the cycle transformation at v_1. Hence, the hypotheses of Theorem 6.19 are satisfied, and so the group generated by $\gamma_1, \ldots, \gamma_4$ is a discrete subgroup of $\text{Möb}(\mathbb{H})$.

Further Reading

As befits a venerable and much studied topic, there is an extensive literature on hyperbolic geometry. The books listed in this chapter form a far from complete list, but they do give you, the interested reader, a place to continue your exploration of hyperbolic geometry.

For the history of the subject, there are the books of Rosenfeld [22], Greenberg [10], and Bonola [4]. Also of interest are the translations by Stillwell [26] of some of the original papers of Beltrami, Klein, and Poincaré, all of whom were instrumental in the development of hyperbolic geometry.

A far from complete list of books discussing various aspects of the hyperbolic geometry of the plane, listed in no particular order, are the books of Trudeau [28], Stahl [24], Rédei [21], Wylie [29], Iversen [13], Coxeter [6], Kelly and Matthews [16], Thurston [27], Fenchel [8], and Pedoe [20], as well as the articles of Beardon [3] and Helmholz [11].

For a more detailed discussion of the properties and implications of discreteness for subgroups of $\text{Möb}(\mathbb{H})$, there are the books of Beardon [2], Katok [15], and the article of Stillwell [25]. There is a very detailed discussion of Poincaré's polygon theorem in the article of Epstein and Petronio [7].

There is an enormous amount of information about hyperbolic geometry available on the World Wide Web as well. At the time of the writing of this book, one very good place to begin searching is the Geometry Center at

http : //www.geom.umn.edu

which has interactive applications, archive materials, pictures, and instructional materials. There is also the Non-Euclid program of Joel Castellanos at

http : //math.rice.edu/ ~ joel/NonEuclid

which is an interactive Java application for exploring the Poincaré disc model of the hyperbolic plane.

Also, there are the works referred to at various points in the text, for further information on specific topics which are not covered in detail in this book.

References

[1] W. Abikoff, 'The uniformization theorem', *Amer. Math. Monthly* **88** (1981), 574–592.

[2] A. F. Beardon, *The Geometry of Discrete Groups*, Graduate Texts in Mathematics, Springer-Verlag, New York, 1983.

[3] A. F. Beardon, 'An introduction to hyperbolic geometry', in *Ergodic Theory, Symbolic Dynamics, and Hyperbolic Spaces*, edited by T. Bedford, M. Keane, and C. Series, Oxford University Press, Oxford, 1991, 1–34.

[4] R. Bonola, *Non-Euclidean geometry*, Dover Publications Inc., New York, 1955.

[5] Yu. D. Burago and V. A. Zalgaller, *Geometric Inequalities*, *Grundlehren der Mathematischen Wissenschaften* 285, Springer-Verlag, New York, 1988.

[6] H. S. M. Coxeter, *Non-Euclidean geometry*, *Mathematical Expositions* 2, University of Toronto Press, Toronto, 1978.

[7] D. B. A. Epstein and C. Petronio, 'An exposition of Poincaré's polyhedron theorem', *Enseign. Math.* (2) **40**, 1994, 113–170.

[8] W. Fenchel, *Elementary geometry in hyperbolic space*, *de Gruyter Studies in Mathematics* 11, Walter de Gruyter, New York, 1989.

[9] L. R. Ford, *Automorphic functions*, Chelsea, New York, 1951.

[10] M. J. Greenberg, *Euclidean and Non-Euclidean Geometries*, W. H. Freeman and Co., New York, 1993.

[11] H. von Helmholz, 'On the Origin and Significance of the Geometrical Axioms', in *The World of Mathematics*, volume 1, edited by J. R. Newman, Simon and Schuster, New York, 1956, 647–668.

[12] I. N. Herstein, *Abstract Algebra*, Prentice Hall Inc., Upper Saddle River, NJ, 1996.

[13] B. Iversen, *Hyperbolic geometry*, *London Mathematical Society Student Texts* 25, Cambridge University Press, Cambridge, 1992.

[14] G. A. Jones and D. Singerman, *Complex functions, an algebraic and geometric viewpoint*, Cambridge University Press, Cambridge, 1987.

[15] S. Katok, Fuchsian groups, Chicago Lectures in Mathematics, University of Chicago Press, Chicago, IL, 1992.

[16] P. J. Kelly and G. Matthews, *The non-Euclidean, hyperbolic plane, Universitext*, Springer-Verlag, New York, 1981.

[17] J. L. Locher, editor, *M. C. Escher, his life and complete graphic work*, Abradale Press, New York, 1992.

[18] B. Maskit, *Kleinian groups*, Springer-Verlag, Berlin, 1988.

[19] J. Munkres, *Topology, a first course*, Prentice-Hall, Inc., Englewood Cliffs, NJ, 1975.

[20] D. Pedoe, *Geometry - a comprehensive course*, Dover Publications, New York, 1988.

[21] L. Rédei, *Foundation of Euclidean and non-Euclidean geometries according to F. Klein, International Series of Monographs in Pure and Applied Mathematics* 97, Pergamon Press, Oxford, 1968.

[22] B. A. Rosenfeld, *A history of non-Euclidean geometry*, Springer-Verlag, New York, 1988.

[23] D. Schattschneider, *Visions of Symmetry*, W. H. Freeman and Company, New York, 1990.

[24] S. Stahl, *The Poincaré Half-Plane*, Jones and Bartlett, Boston, 1993.

[25] J. Stillwell, *Poincaré, geometry and topology*, in *Henri Poincaré: science et philosophie*, Akademie Verlag, Berlin, 1996, 231–240.

[26] J. Stillwell, *Sources of Hyperbolic Geometry*, History of Mathematics, volume 10, American Mathematical Society, Providence, RI, 1996.

[27] W. P. Thurston, *Three-Dimensional Geometry and Topology*, Princeton University Press, Princeton, 1997.

[28] R. J. Trudeau, *The non-Euclidean revolution*, Birkhäuser Boston, Boston, MA, 1987.

[29] C. R. Wylie, Jr., *Foundations of geometry*, McGraw-Hill Book Co., New York, 1964.

Notation

The purpose of this chapter is to provide a list of the various bits of notation which appear throughout the book. The chapter or section in which the notation first appears is given in brackets.

$\mathbb{R}$	the real numbers [1.1]
$\mathbb{C}$	the complex numbers [1.1]
$\mathrm{Re}(z)$	$= x$, the real part of the complex number $z = x + iy$ [1.1]
$\mathrm{Im}(z)$	$= y$, the imaginary part of $z = x + iy$ [1.1]
$\lvert z\rvert$	$= \sqrt{(\mathrm{Re}(z))^2 + (\mathrm{Im}(z))^2}$, the norm or modulus of z [1.1]
$\mathbb{H}$	$= \{z \in \mathbb{C} \mid \mathrm{Im}(z) > 0\}$, the upper half plane in $\mathbb{C}$ [1.1]
$\mathbb{N}$	$= \{1, 2, 3, \ldots\}$, the natural numbers [1.2]
$\mathbb{Q}$	the rational numbers [1.2]
$\overline{\mathbb{C}}$	$= \mathbb{C} \cup \{\infty\}$, the Riemann sphere [1.2]
$\mathbb{Z}$	$= \{0, \pm 1, \pm 2, \ldots\}$, the integers [1.2]
$\mathbb{D}$	$= \{z \in \mathbb{C} \mid \lvert z\rvert < 1\}$, the open unit disc in $\mathbb{C}$ [1.3, 4.1]
$\mathbb{S}^1$	$= \{z \in \mathbb{C} \mid \lvert z\rvert = 1\}$, the unit circle in $\mathbb{C}$ [1.1]
$\mathbb{S}^2$	the unit sphere in $\mathbb{R}^3$, [1.2]
$U_\varepsilon(z)$	the open disc of radius ε and centre z [1.2, 5.3, 3.3]
$U_\varepsilon(\infty)$	the open disc in $\overline{\mathbb{C}}$ of radius ε and centre ∞ [1.2]
$\overline{\mathbb{R}}$	$= \mathbb{R} \cup \{\infty\}$, the extended real line [1.2]
$\overline{X}$	the closure of a set X in $\overline{\mathbb{C}}$ [1.2]
$\mathrm{Homeo}(\overline{\mathbb{C}})$	group of homeomorphisms of $\overline{\mathbb{C}}$ [1.2]
$\mathrm{Homeo}^C(\overline{\mathbb{C}})$	set of circle preserving homeomorphisms of $\overline{\mathbb{C}}$ [2.1]
$\mathrm{M\ddot{o}b}^+$	group of Möbius transformations [2.1]
$[z_1, z_2; z_3, z_4]$	the cross ratio [2.3]

Möb	the general Möbius group [2.6]
$\text{angle}(C_1, C_2)$	angle between curves [2.7]
$\text{Möb}(\mathbb{H})$	subgroup of Möb preserving $\mathbb{H}$ [2.8]
$\text{Möb}(\overline{\mathbb{R}})$	subgroup of Möb preserving $\overline{\mathbb{R}}$ [2.8]
$\text{Möb}(A)$	subgroup of Möb preserving the circle A [2.8]
$\text{Möb}(\mathbb{D})$	subgroup of Möb preserving $\mathbb{D}$ [2.8]
$\text{Möb}^+(\mathbb{H})$	subgroup of Möb^+ preserving $\mathbb{H}$ [2.8]
$\min(x, y)$	the minimum of the two numbers x and y [1.2]
$\text{bij}(X)$	the group of bijections from X to X [2.2]
$\det(m)$	the determinant of the Möbius transformation m [2.5]
$\tau(m)$	the square of the trace of the Möbius transformation m [2.5]
$\text{GL}_2(\mathbb{C})$	2×2 invertible matrices over $\mathbb{C}$ [2.5]
$\text{SL}_2(\mathbb{C})$	2×2 matrices with determinant 1 over $\mathbb{C}$ [2.5]
$\rho(z)\,\lvert dz\rvert$	general element of arc-length [3.1]
$\text{length}_r ho(f)$	length with respect to $\rho(z)\,\lvert dz\rvert$ [3.1]
$\text{length}_{\mathbb{H}}(f)$	hyperbolic length in $\mathbb{H}$ [3.2]
(X, d)	a metric space [3.3]
$\Gamma[x, y]$	paths in $\mathbb{H}$ from x to y [3.3]
$\text{d}_{\mathbb{H}}(x, y)$	hyperbolic distance between points in $\mathbb{H}$ [3.4]
$(\mathbb{H}, \text{d}_{\mathbb{H}})$	the upper half-plane as a metric space [3.4]
$\text{Isom}(\mathbb{H})$	group of isometries of $(\mathbb{H}, \text{d}_{\mathbb{H}})$ [3.6]
$\text{d}_{\mathbb{H}}(X, Y)$	hyperbolic distance between sets in $\mathbb{H}$ [3.7]
$\text{Möb}(\mathbb{D})$	subgroup of Möb preserving $\mathbb{D}$ [4.1]
$\text{Möb}^+(\mathbb{D})$	subgroup of Möb^+ preserving $\mathbb{D}$ [4.1]
$\text{length}_{\mathbb{D}}(f)$	hyperbolic length in $\mathbb{D}$ [4.1]
$\Theta[x, y]$	paths in $\mathbb{D}$ from x to y [4.1]
$\text{d}_{\mathbb{D}}(x, y)$	hyperbolic distance between points in $\mathbb{D}$ [4.1]
$(\mathbb{D}, \text{d}_{\mathbb{D}})$	the Poincare disc as a metric space [4.1]
$\text{curv}(z)$	curvature [4.2]
$\text{conv}(X)$	the convex hull of X [5.1]
$\text{area}_{\mathbb{H}}(X)$	hyperbolic area in $\mathbb{H}$ [5.4]
$\text{area}_{\mathbb{D}}(X)$	hyperbolic area in $\mathbb{D}$ [5.4]
$\Gamma(z)$	the orbit of z under Γ [6.2]
$\text{PSL}_2(\mathbb{Z})$	the modular group [6.2]
Γ_x	stabilizer of x in Γ [6.2]
$z \sim_\Gamma w$	equivalent under the action of Γ [6.3]
$\text{stab}_\Gamma(X)$	stabilizer of X in Γ [6.3]
$\Gamma(X)$	union of translates of X under Γ [6.3]

$D_\Gamma(z_0)$	the Dirichlet polygon [6.4]
$\mathrm{int}(X)$	the interior of a set X in $\mathbb{H}$ [6.4]
$\mathrm{sum}(v_0)$	angle sum of cycle transformation [6.5]

Index